Model-Based Control of Particulate Processes

Particle Technology Series

Series Editor

Professor Brian Scarlett
Technical University of Delft

The Kluwer Particle Technology Series of books is the successor to the Chapman and Hall Powder Technology Series. The aims are the same, the scope is wider. The particles involved may be solid or they may be droplets. The size range may be granular, powder or nano-scale. The accent may be on materials or on equipment, it may be practical or theoretical. Each book can add one brick to a fascinating and vital technology. Such a vast field cannot be covered by a carefully organised tome or encyclopaedia. Any author who has a view or experience to contribute is welcome. The subject of particle technology is set to enjoy its golden times at the start of the new millennium and I expect that the growth of this series of books will reflect that trend.

The titles published in this series are listed at the end of this volume.

Model-Based Control of Particulate Processes

by

PANAGIOTIS D. CHRISTOFIDES

Department of Chemical Engineering,
University of California,
Los Angeles, CA, U.S.A.

KLUWER ACADEMIC PUBLISHERS
DORDRECHT / BOSTON / LONDON

A C.I.P. Catalogue record for this book is available from the Library of Congress.

ISBN 1-4020-0936-4

Published by Kluwer Academic Publishers,
P.O. Box 17, 3300 AA Dordrecht, The Netherlands.

Sold and distributed in North, Central and South America
by Kluwer Academic Publishers,
101 Philip Drive, Norwell, MA 02061, U.S.A.

In all other countries, sold and distributed
by Kluwer Academic Publishers,
P.O. Box 322, 3300 AH Dordrecht, The Netherlands.

Printed on acid-free paper

Printed in the Netherlands.

Contents

List of Figures

List of Tables

Preface

Particulate processes are characterized by the co-presence of a continuous phase and a dispersed (particulate) phase, and are widely used in industry for the manufacturing of many high-value products. Examples include the crystallization of proteins for pharmaceutical applications, the emulsion polymerization reactors for the production of latex, the aerosol synthesis of titania powder used in the production of white pigments, and the thermal spray processing of nanostructured coatings. It is now well understood that the physico-chemical and mechanical properties of materials made with particulates depend heavily on the characteristics of the corresponding particle size distribution. This fact, together with recent advances in dynamics of infinite-dimensional systems and nonlinear control theory, has motivated extensive research on model-based control of particulate processes using population balances to achieve tight control of particle size distributions.

This book - the first of its kind - presents general methods for the synthesis of nonlinear, robust and constrained feedback controllers for broad classes of particulate process models and illustrates their applications to industrially-important crystallization, aerosol and thermal spray processes. The controllers use a finite number of measurement sensors and control actuators to achieve stabilization of the closed-loop system, output tracking, attenuation of the effect of model uncertainty and handling of actuator saturation.

Specifically, our attention focuses on nonlinear order reduction and nonlinear, robust and constrained control of spatially-homogeneous particulate processes, and nonlinear control of spatially-distributed particulate processes. We use geometric and Lyapunov-based control techniques to synthesize nonlinear, robust and bounded controllers that use a finite number of measurement sensors and control actuators to achieve stabilization of the closed-loop system, output tracking, attenuation of the effect of model uncertainty and handling of actuator saturation. The methods are applied to continuous and batch crystallization processes, a titania aerosol reactor and a thermal spray process to regulate product size distribution. The resulting benefits in closed-loop performance, robustness and actuator saturation handling compared to other techniques for control of particulate processes are demonstrated through computer simulations. The book includes comparisons of the methods followed for controller synthesis with other approaches and discussions of practical implementation issues that can help researchers and engineers understand the development and application of the methods in greater depth.

The book assumes a basic knowledge about population balances and nonlinear control and is intended for researchers, graduate students and process control engineers interested in particulate processes, nonlinear control and process control applications.

All the research that led to this book was carried out at the Department of Chemical Engineering at the University of California, Los Angles. In addition to my work, my graduate students, Timothy Chiu, Nael El-Farra, Ashish Kalani and Mingheng Li contributed in research results included in the book and the preparation of the final manuscript. I would like to thank them for their hard work and contributions.

I would also like to thank all the other people who contributed in some way on this project. In particular, I would like to thank my faculty colleagues at UCLA for creating a pleasant working environment, the staff of Kluwer Academic for excellent cooperation, and the Petroleum Research Fund, the United States National Science Foundation, and the United States Office of Naval Research for financial support. Last, but not least, I would like to express my deepest gratitude to my wife and my parents for their dedication, encouragement and support over the course of this project. I dedicate this book to them.

PANAGIOTIS D. CHRISTOFIDES

Chapter 1

INTRODUCTION

Particulate processes (also known as dispersed-phase processes) are characterized by material domains that are comprised of a continuous phase and a dispersed phase and are essential in making many high-value industrial products. Examples include the crystallization of proteins for pharmaceutical applications, the emulsion polymerization reactors for the production of latex, the aerosol synthesis of titania powder used in the production of white pigments, and the thermal spray processing of nanostructured coatings. It is now well understood that the physico-chemical and mechanical properties of materials made with particulates depend heavily on the characteristics of the corresponding particle-size distribution (PSD) (for example, a nearly monodisperse PSD is required for titania pigments to obtain the maximum hiding powder per unit mass). Therefore, the problem of synthesizing and implementing high-performance model-based feedback control systems on particulate processes to achieve PSDs with desired characteristics has significant industrial value. On the other hand, recent developments in measurement technology allow the accurate and fast on-line measurement of key process variables including PSDs (see, for example, the article by Rawlings et al., 1993 for an excellent review of the available measurement technology), which can then be used for feedback control purposes.

Population balances have provided a natural framework for the mathematical modeling of PSDs (see, for example, the tutorial article by Hulburt and Katz, 1964 and the review article by Ramkrishna, 1985), and have been successfully used to describe PSDs in emulsion polymerization reactors (e.g., Dimitratos et al., 1994), crystallizers (e.g., Rawlings et al., 1993) and aerosol reactors (e.g., Friedlander, 1977; William and Loyalka, 1991) and cell cultures (e.g., Daoutidis and Henson, 2001). Application of population balances to particulate processes typically leads to systems of nonlinear partial integro-differential equations that describe the rate of change of the PSD. The population balances are coupled with the material and energy balances that describe the rate of change of the state variables of the continuous phase (these are usually systems of nonlinear differential equations, which include integrals over the entire particle-size spectrum), leading to complete particulate process models.

The nonlinear (owing to complex growth, nucleation, agglomeration and breakage mechanisms, and their Arrhenius dependence on temperature) and distributed nature of population balances has motivated extensive research on the development of efficient numerical methods for the accurate computation of their solution. Examples of solution methods for continuous population balances include the method of self-preserving distributions (Friedlander, 1977), the method of weighted residuals (e.g., Ramkrishna, 1985; Gelbard and Seinfeld, 1978), the sectional method (e.g., Gelbard et al., 1980; Landgrebe and Pratsinis, 1990), and discretization via fixed/moving pivot techniques (Kumar and Ramkrishna, 1996a; Kumar and Ramkrishna, 1996b), while methods for the solution of discretized population balances have been proposed by Hounslow, 1990 and Hill and Ng, 1996. An excellent review of results in this area can be found in the review article by Ramkrishna, 1985 (see also Daoutidis and Henson, 2001). The ability to accurately solve population balance models motivated numerous research studies on the dynamics of particulate processes. These studies confirmed the existence of a wide range of complex static and dynamic phenomena including multiple steady-states and sustained oscillations (see, for example, Lei et al., 1971; Jerauld et al., 1983 and Rawlings and Ray, 1987a; Rawlings and Ray, 1987b for a theoretical analysis of oscillatory behavior in crystallizers and emulsion polymerization reactors, respectively), which had been previously observed in experimental studies (see the classic book by Randolph and Larson, 1988 for results and references). The highly nonlinear and oscillatory behavior of many particulate processes implies the need to implement nonlinear model-based feedback controllers in order to ensure a stable and efficient operation.

In spite of the rich literature on population balance modeling, numerical solution, and dynamical analysis of particulate processes, research on model-based control of such processes has been very limited. Specifically, previous efforts have mainly focused on the understanding of fundamental control-theoretic properties (controllability and observability) of population balance models (Semino and Ray, 1995a) and the application of conventional control schemes (such as proportional-integral and proportional-integral-derivative control, self-tuning control) to crystallizers and emulsion polymerization processes (see, for example, Semino and Ray, 1995b; Rohani and Bourne, 1990; Dimitratos et al., 1994 and the references therein). Notable exceptions on model-based control of particulate processes include an optimization-based control method which was developed by Eaton and Rawlings, 1990, and successfully applied to a batch crystallization process, as well as nonlinear state feedback control of a cell culture by Kurtz et al., 1998. The main difficulty in synthesizing nonlinear model-based feedback controllers for particulate processes is the distributed parameter nature of the population balance models which does not allow their direct use for the synthesis of low-order (and therefore, practically implementable) nonlinear output feedback controllers. Furthermore, a direct application of the aforementioned solution methods to derive finite dimensional approximations of the population balance models may lead to ODE systems of very high order, which are inappropriate for the synthesis of low-order controllers.

In addition to being highly nonlinear and infinite dimensional, the population balance models of most particulate processes are uncertain. Typical sources of model uncertainty include unknown or partially known time-varying process parameters, exogenous disturbances and unmodeled dynamics. It is well-known that the presence of uncertain variables and unmodeled dynamics, if not taken into account in the controller design, may lead to severe deterioration of the nominal closed-loop performance or even to closed-loop instability. Research on robust control of nonlinear distributed chemical processes with uncertainty has mainly focused on transport-reaction processes described by nonlinear partial differential equations (PDEs). In this area, important contributions include the development of Lyapunov-based robust control methods for hyperbolic (Christofides and Daoutidis, 1998) and parabolic (Ydstie and Alonso, 1997; Christofides, 1998) PDEs; the reader may also refer to the recent book by Christofides, 2001 for results and references in this area. An alternative approach for the design of controllers for PDE systems with time-invariant uncertain variables involves the use of adaptive control methods (e.g., Byrnes, 1987; Wen and Balas, 1989; Balas, 1995). Despite this progress, at this stage, there is no general framework for the synthesis of practically-implementable nonlinear feedback controllers for particulate processes that allow attaining desired particle size distributions in the presence of significant model uncertainty.

Another unresolved issue on control of particulate processes is the ability of the controller to deal with constraints in the capacity of the control actuators. This is an important limitation of these methods, especially in view of the fact that the capacity of control actuators used to regulate particulate processes is almost always limited. Such limitations may arise naturally due to the finite capacity of control actuators (e.g., bounds on magnitude of the opening of valves) or may be imposed on the manipulated input to insure safe process operation, meet environmental regulations, or maintain desired product quality specifications. Input constraints restrict our ability to freely modify the dynamic behavior of particulate processes; and the ill-effects due to actuator constraints manifest themselves, for example, in the form of sluggishness of response and loss of stability. Additional problems that arise in the case of dynamic controllers include undesired oscillations and overshoots, a phenomenon usually referred to as "windup".

Recognition of the detrimental effects of input constraints on the stability and performance of chemical processes in general has motivated many recent studies on the dynamics and control of chemical processes subject to input constraints. Notable contributions in this regard include controller design and stability analysis within the model predictive control framework (see, for example, Rao and Rawlings, 1999; Scokaert and Rawlings, 1999), the design of "anti-windup" schemes in order to prevent excessive performance deterioration of an already designed controller when the input saturates (e.g., Kothare et al., 1994; Kapoor et al., 1998), and the design of bounded feedback control laws for various classes of nonlinear finite-dimensional systems (Alvarez et al., 1991; El-Farra and Christofides, 2001a) and, recently, for parabolic PDE systems (El-Farra et al., 2002). While the control of nonlinear finite-dimensional systems with input constraints has received significant attention, there are no results on model-based control of particulate processes with input constraints.

Motivated by the lack of population balance-based control methods for particulate processes that can deal with the practical issues of nonlinear behavior, model uncertainty and actuator saturation, and the need to achieve tight size distribution control in many particulate processes, the objectives of the present book are:

- to present general nonlinear, robust and constrained control methods for particulate processes based on population balance models, and
- to illustrate the application of the proposed methods to size distribution control in continuous and batch crystallization, and aerosol and thermal spray processes and document their effectiveness and advantages with respect to conventional control methods.

The rest of the book is structured as follows.

Chapter 2 proposes a general method for the synthesis of practically implementable nonlinear model-based feedback controllers that enforce the desired stability and performance specifications (such as PSDs with desired total mass and mean particle size) in the closed-loop system. A a model reduction procedure based on a combination of the method of weighted residuals and the concept of approximate inertial manifold is proposed and used for the construction of low-order ODE systems that accurately reproduce the dominant dynamics of the particulate process. These ODE systems are subsequently used for the synthesis of nonlinear low-order output feedback controllers that enforce exponential stability in the closed-loop system and achieve a desired particle size distribution. Precise closed-loop stability conditions are given and controller implementation issues are addressed.

Chapter 3 shows two applications of the nonlinear control method proposed in Chapter 2 to continuous and batch crystallization systems. Low-dimensional approximations of the population balances are derived and used for nonlinear control design. The performance of the nonlinear controllers is successfully tested through simulations using the detailed population balance models and is shown to be superior to the one of linear control schemes.

Chapter 4 proposed a general method for the synthesis of practically-implementable robust nonlinear controllers that explicitly handle time-varying uncertain variables (e.g., unknown process parameters and disturbances) and unmodeled dynamics (e.g., fast actuator and sensor dynamics not included in the process model). The robust nonlinear controllers are synthesized via Lyapunov's direct method and are shown to enforce stability in the closed-loop system and attenuation of the effect of uncertain variables on the outputs, and achieve particle size distributions with desired characteristics. The method is successfully applied to a continuous crystallizer with fines trap and is shown to outperform a nonlinear controller design which does not account for the effect of time-varying uncertainty.

Chapter 5 presents a general methodology for the analysis and control of constrained particulate processes modeled by population balance equations. Using finite-dimensional ODE approximations of the population balances, we characterize the limitations imposed by input constraints on our ability to modify the dynamics of the particulate process and provide an explicit characterization of the set of admissible set-points that can be achieved in the presence of constraints. This information, together with the derived ODE systems, is then used as the basis for the synthesis of practically-implementable nonlinear bounded output feedback controllers that enforce exponential stability in the closed-loop system and achieve particle size distributions with desired characteristics in the presence of input constraints. The approach is used to control a continuous crystallizer with actuator saturation.

Chapter 6 presents a nonlinear feedback control method for spatially-distributed aerosol processes for which the manipulated inputs, the control objectives and the measurements are distributed in space. Initially, a general population balance model (a nonlinear partial integro-differential equation) which describes aerosol processes with simultaneous chemical reaction, nucleation, condensation, coagulation and convective transport is presented. Then, under the assumption of lognormal aerosol size distribution, the method of moments is employed to reduce the population balance model into a set of first-order hyperbolic PDEs which accurately describes the spatio-temporal evolution of the three leading moments needed to exactly characterize the aerosol size distribution. This hyperbolic PDE system is then used as the basis for the synthesis of nonlinear distributed output feedback controllers that use process measurements to achieve an aerosol size distribution with desired characteristics (e.g. geometric average particle volume). The performance of the proposed control method is successfully tested through simulations on a typical aerosol process.

Chapter 7 focuses on modeling and control of an aerosol flow reactor used to produce titania powder by gas phase oxidation of titanium tetrachloride. A fundamental mathematical model is derived for the process which describes the spatio-temporal evolution of the three leading moments of the aerosol size distribution, as well as the evolution of the concentrations of the species and temperature of the continuous phase. The model accounts for simultaneous chemical reaction, nucleation, Brownian and turbulent coagulation and convective transport and comprises of eight nonlinear first-order hyperbolic partial differential equations (PDEs). Then, the process model is used to synthesize a nonlinear output feedback controller, using the method presented in chapter 6, which manipulates the temperature of the reactor wall to achieve an aerosol size distribution in the outlet of the reactor with desired geometric average particle diameter. The performance and robustness of the nonlinear controller is successfully tested through computer simulations.

Chapter 8 presents a formulation of the control problem and a feedback control system for the high-velocity oxygen-fuel (HVOF) thermal spray process, which explicitly accounts for the effect of powder size distribution. Based on the predictions of a fundamental model which is capable of capturing the essential features of the HVOF process, and available experimental data, the control problem is formulated as the one of regulating volume-based averages of the temperature and velocity of the particles at the point of impact on substrate by manipulating the oxygen/fuel ratio and the combustion chamber pressure, respectively. A feedback control system is developed and applied to the detailed model of the process. Closed-loop simulations show that the volume-based average particle velocity and temperature at the point of impact on substrate reach the desired values in a short time. It is also shown that the proposed formulation of the control problem (which accounts for the effect of powder size distribution) leads to a solution of the control problem that is superior (with respect to the achievement of the desired control objectives) to a solution that assumes a monodisperse powder size distribution.

The proofs of all the theoretical results given in the main body of the book are included in the appendix.

Chapter 2

NONLINEAR MODEL REDUCTION AND CONTROL OF PARTICULATE PROCESSES

1. Introduction

This chapter focuses on nonlinear model reduction and feedback control of particulate processes for which the effect of spatial variations can be safely neglected. Using population balances, such processes can be adequately described by a class of nonlinear partial integro-differential equation systems. The objective is to develop a general and rigorous method for the derivation of low-order approximations of the distributed parameter process model and the synthesis of practically implementable (i.e., low-order) nonlinear model-based feedback controllers that enforce the desired stability and performance specifications (such as PSDs with desired total mass and mean particle size) in the closed-loop system.

This chapter is structured as follows: Section 2 presents a general class of nonlinear partial integro-differential equation systems which describes the majority of spatially-homogeneous particulate processes and uses a crystallizer example to motivate the proposed approach for control of particulate processes. Section 3 proposes a model reduction procedure based on a combination of the method of weighted residuals and the concept of approximate inertial manifold for the construction of low-order ODE systems that accurately reproduce the dominant dynamics of the particulate process. These ODE systems are subsequently used for the synthesis of nonlinear low-order output feedback controllers that enforce exponential stability in the closed-loop system and achieve a desired particle size distribution. Precise closed-loop stability conditions are given and controller implementation issues are addressed. In Chapter 3, the proposed control method is applied to continuous and batch crystallization systems and its performance and robustness properties are tested through simulations. The results of this chapter were first presented in Christofides and Chiu, 1997; Chiu and Christofides, 1999a.

2. Modeling and Dynamics of Particulate Processes

2.1 Particulate Process Model

We focus on spatially homogeneous (well-mixed) particulate processes with simultaneous particle growth, nucleation, agglomeration, and breakage and consider the case of a single internal particle coordinate which, for the sake of exposition, is assumed to be the particle-size. Applying a dynamic material balance on the number of particles of size r to $r + dr$ (population balance), we obtain the following general nonlinear partial integro-differential equation which describes the rate of change of the PSD, $n(r, t)$:

$$\frac{\partial n}{\partial t} = -\frac{\partial (G(x, r)n)}{\partial r} + w(n, x, r) \tag{2.1}$$

where $n(r, t) \in L_2[0, r_{max})$ ($L_2[0, r_{max})$ is the Hilbert space of continuous functions defined on the interval $[0, r_{max})$), $r \in [0, r_{max}]$ is the particle size, and r_{max} is the maximum particle size (which may be infinity), t is the time and $x \in \mathbb{R}^n$ is the vector of state variables which describe properties of the continuous phase (for example solute concentration, temperature, and pH in a crystallizer); see Eq.2.2 below for the system that describes the dynamics of x. $G(x, r)$ and $w(n, x, r)$ are nonlinear scalar functions whose physical meaning can be explained as follows: $G(x, r)$ accounts for particle growth through condensation and is usually referred to as growth rate. It usually depends on the concentrations of the various species present in the continuous phase, the temperature of the process, and the particle size. On the other hand, $w(n, x, r)$ represents the net rate of introduction of new particles into the system. It includes all the means by which particles appear or disappear within the system including particle agglomeration (merging of two particles into one), breakage (division of one particle to two), as well as, nucleation of particles of size $r \geq 0$ and particle feed and removal.

The rate of change of the continuous-phase variables x can be derived by a direct application of mass and energy balances to the continuous phase and is given by a nonlinear integro-differential equation system of the general form:

$$\dot{x} = f(x) + g(x)u(t) + A\int_0^{r_{max}} a(n, r, x)dr \tag{2.2}$$

where $f(x), a(n, r, x)$ are nonlinear vector functions, $g(x)$ is a nonlinear matrix function, A is a constant matrix and $u(t) = [u_1 \ u_2 \ \cdots \ u_m] \in$

$\mathbb{R}^m$ is the vector of manipulated inputs. The term $A\int_0^{r_{max}} a(n,r,x)dr$ accounts for mass and heat transfer from the continuous phase to all the particles in the population.

We define a vector of controlled outputs to express the various control objectives (such as regulation of total number of particles, mean particle size, temperature, pH, and so on) as:

$$y_i(t) \;=\; h_i\Big(\int_0^{r_{max}} c_\kappa(r)n(r,t)dr, x\Big), \;\; i=1,\ldots,m, \;\; \kappa=1,\ldots,l \tag{2.3}$$

where $y_i(t)$ is the ith controlled output, $h_i(\int_0^{r_{max}} c_\kappa(r)n(r,t)dr, x)$ is a nonlinear smooth scalar function of its arguments, and $c_\kappa(r)$ is a known smooth function of r which depends on the desired performance specifications. To simplify the notation of the theoretical development, we will not consider measured outputs separately from controlled outputs, which means that we need to assume the availability of on-line measurements of the controlled outputs $y_i(t)$.

Throughout the thesis, we will use the inner product and norm in $L_2[0, r_{max})$, which are defined, respectively, as:

$$(\phi_1, \phi_2) \;=\; \int_0^{r_{max}} \phi_1(z)\phi_2(z)dz, \quad ||\phi_1||_2 = (\phi_1, \phi_1)^{\frac{1}{2}} \tag{2.4}$$

where ϕ_1, ϕ_2 are two elements of $L_2[0, r_{max})$. Furthermore, the order of magnitude and Lie derivative notations will be needed in our development. In particular, $\delta(\epsilon) = O(\epsilon)$ if there exist positive real numbers k_1 and k_2 such that $|\delta(\epsilon)| \leq k_1|\epsilon|$, $\forall$ $|\epsilon| < k_2$, and $L_f\bar{h}$ denotes the standard Lie derivative of a scalar function $\bar{h}(x)$ with respect to the vector function $f(x)$, $L_f^k\bar{h}$ denotes the k-th order Lie derivative and $L_gL_f^{k-1}\bar{h}$ denotes the mixed Lie derivative where $g(x)$ is a vector function.

Remark 2.1: Referring to the general mathematical model of Eqs.2.1-2.2, the following remarks are in order: a) the particle-size distribution function $n(r,t)$ is assumed to be a sufficiently smooth function of its arguments (that is, $n(r,t)$ and its partial derivatives with respect to r and t, up to a desired order, are continuous functions); this is a reasonable assumption for large size distributions, even though particles are discrete and their number is integer-valued, b) a single internal coordinate (particle size) is considered; this is motivated by the majority of industrial particulate process control problems where the central objective is to produce particulates with a desired particle size distribution, c) the particles are assumed to be small enough so that the environment in which they are dispersed can be adequately described by a local value of its state vector, d) the vector of manipulated inputs $u(t)$ is a lumped variable (that is, it only depends on time; for example, the inlet concentration of solute in a crystallizer) and the controlled outputs $y_i(t)$ are taken to be nonlinear functions of x and appropriate weighted averages of n (examples of controlled outputs included in this formulation are total mass of particles, mean particle size, solute concentration, process temperature, and pH); both choices are typical in most practical applications as illustrated by the continuous and batch crystallization processes studied in Chapter 3.

Remark 2.2: The nonlinear model reduction and controller synthesis results that will be presented in this thesis can be generalized to particulate processes, which include manipulated inputs in the population balance (such as manipulation of fines destruction rate; see Lei et al., 1971; Randolph et al., 1987, and Remark 2.6 below) with the following state-space description:

$$\begin{aligned}
\frac{\partial n}{\partial t} &= -\frac{\partial(G(x,r)n)}{\partial r} + w(n,x,r) + g_1(n,x,r)u(t) \\
\dot{x} &= f(x) + g_2(x, \int_0^{r_{max}} a_1(n,r,x)dr)u(t) \\
&\quad + g_3(x, \int_0^{r_{max}} a_2(n,r,x)dr) \\
y_i(t) &= h_i(\int_0^{r_{max}} c_\kappa(r)n(r,t)dr, x), \quad i = 1,\ldots,m, \quad \kappa = 1,\ldots,l
\end{aligned} \tag{2.5}$$

where $g_1(n,x,r), g_3(x, \int_0^{r_{max}} a_2(n,r,x)dr), a_1(n,r,x), a_2(n,r,x)$ are nonlinear vector functions and $g_2(x, \int_0^{r_{max}} a_1(n,r,x)dr)$ is a nonlinear matrix function.

2.2 Dynamics of Particulate Processes: A Continuous Crystallizer

Crystallization is a particulate process which is widely used in industry for the production of many products including fertilizers, proteins, and pesticides. The fact that the shape of the crystal-size distribution influences significantly the necessary liquid-solid separation, as well as the properties of the product, implies that crystallization requires a population balance in order to be accurately described, analyzed, and controlled. Crystallizers typically exhibit highly oscillatory behavior which suggests the use of feedback control to ensure stable operation and attain a crystal-size distribution with desired characteristics.

A continuous crystallizer (Figure 2.1), which exhibits highly oscilla-

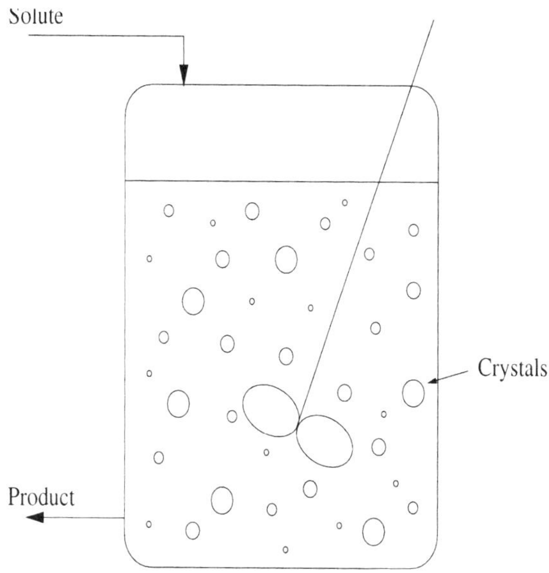

Figure 2.1. A continuous crystallizer.

tory behavior, is considered, and it will be shown that its dominant dynamics can be accurately described by a small set of ODEs; this finding implies that the dominant dynamics of the process are characterized by a small number of degrees of freedom (low-dimensional) and motivates the proposed approach for nonlinear model-based feedback control of particulate processes. Under the assumptions of isothermal operation, constant volume, mixed suspension, nucleation of crystals of infinitesimal size and mixed product removal, a dynamic model for the crystallizer can be derived from a population balance for the particle phase and a mass balance for the solute concentration, and it is of the following form (Lei et al., 1971; Jerauld et al., 1983):

$$\begin{aligned}\frac{\partial n}{\partial \bar{t}} &= -\frac{\partial (R(\bar{t})n)}{\partial r} - \frac{n}{\tau} + \delta(r-0)Q(\bar{t}) \\ \frac{dc}{d\bar{t}} &= \frac{(c_0-\rho)}{\bar{\epsilon}\tau} + \frac{(\rho-c)}{\tau} + \frac{(\rho-c)}{\bar{\epsilon}}\frac{d\bar{\epsilon}}{d\bar{t}}\end{aligned} \tag{2.6}$$

where $n(r,\bar{t})$ is the number of crystals of radius $r \in [0,\infty)$ at time $\bar{t}$ per unit volume of suspension, τ is the residence time, c is the solute concentration in the crystallizer, c_0 is the solute concentration in the feed, and $\bar{\epsilon} = 1 - \int_0^{\infty} n(r,\bar{t})\frac{4}{3}\pi r^3 dr$ is the volume of liquid per unit volume of suspension. $R(\bar{t})$ is the growth rate, $\delta(r-0)$ is the standard Dirac function and $Q(\bar{t})$ is the nucleation rate. The term $\delta(r-0)Q(\bar{t})$ accounts for the production of crystals of infinitesimal (zero) size via nucleation. $R(\bar{t})$ and $Q(\bar{t})$ are assumed to follow McCabe's law and Volmer's nucleation law, respectively:

$$R(\bar{t}) = k_1(c - c_s), \quad Q(\bar{t}) = \bar{\epsilon} k_2 e^{-\frac{k_3}{\left(\frac{c}{c_s}-1\right)^2}} \tag{2.7}$$

where k_1,k_2,and k_3 are constants and c_s is the concentration of solute at saturation. Using the expressions for $Q(\bar{t})$ and $R(\bar{t})$, the system of Eq.2.6 can be equivalently written as:

$$\begin{aligned} \frac{\partial n}{\partial \bar{t}} &= -k_1(c - c_s)\frac{\partial n}{\partial r} - \frac{n}{\tau} + \delta(r-0)\bar{\epsilon}k_2 e^{-\frac{k_3}{\left(\frac{c}{c_s}-1\right)^2}} \\ \frac{dc}{d\bar{t}} &= \frac{(c_0 - \rho)}{\bar{\epsilon}\tau} + \frac{(\rho - c)}{\tau} + \frac{(\rho - c)}{\bar{\epsilon}}\frac{d\bar{\epsilon}}{d\bar{t}} \end{aligned} \tag{2.8}$$

To study the dynamic behavior of the crystallizer in question, a second-order accurate finite difference scheme with 1,000 discretization points was used to obtain the solution of the system of Eq.2.6 (the values of the process parameters used in the simulations are given in Tables 2.1 and 2.2). Figure 2.2 shows the open-loop oscillatory profiles of the total

c_0	=	1000.0	$kg\ m^{-3}$
c_s	=	980.2	$kg\ m^{-3}$
ρ	=	1770.0	$kg\ m^{-3}$
τ	=	1.0	hr
k_1	=	5.065×10^{-2}	$mm\ m^3\ kg^{-1}\ hr^{-1}$
k_2	=	7.958	$mm^{-3}\ hr^{-1}$
k_3	=	1.217×10^{-3}	

Table 2.1. Parameters for continuous crystallizer model.

crystal concentration and total crystal size, respectively. It is clear that the crystallizer exhibits highly oscillatory behavior, which is the result of the interplay between growth and nucleation caused by the relative nonlinearity of the nucleation rate as compared to growth rate (compare the nonlinear dependence of $Q(\bar{t})$ and $R(\bar{t})$ on c in Eq.2.7). To establish

σ	$=$	$k_1\tau(c_0 - c_s)$	$=$	1.0	mm
Da	$=$	$8\pi\sigma^3 k_2 \tau$	$=$	200.0	
F	$=$	$k_3 c_s^2/(c_0 - c_s)^2$	$=$	3.0	
α	$=$	$(\rho - c_s)/(c_0 - c_s)$	$=$	40.0	

Table 2.2. Dimensionless parameters for continuous crystallizer model.

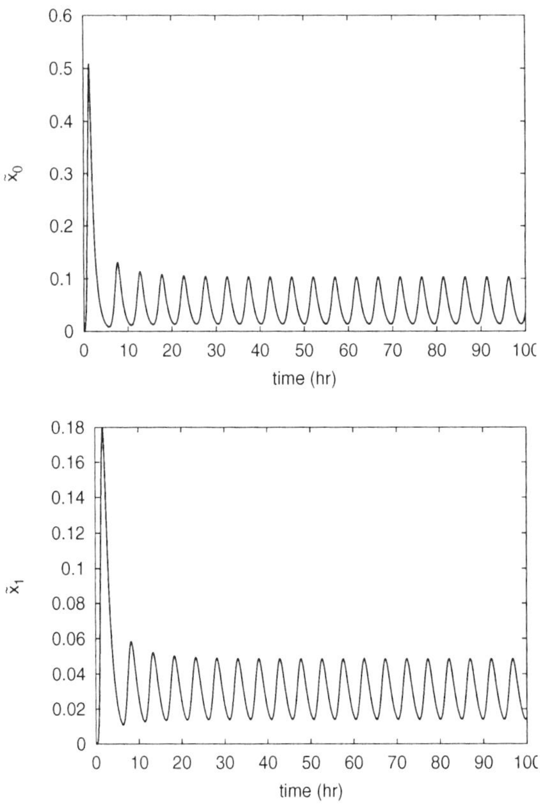

Figure 2.2. Open-loop profiles of crystal concentration (top) and total crystal size (bottom); distributed parameter model.

that the dynamics of the crystallizer are characterized by a small number of degrees of freedom, the method of moments is applied to the system of Eq.2.6 to derive an approximate ODE model. This is possible because the nucleation and growth rates are assumed to be independent of particle size which allows closure of the moment equations (see Eq.2.12 below). It is noted that the method of moments has been extensively used in the past to analyze the dynamics of particulate processes (see, for example, the articles by Hulburt and Katz, 1964 and Pratsinis, 1988).

Defining the νth moment of $n(r,\bar{t})$ as:

$$\mu_\nu = \int_0^\infty r^\nu n(r,\bar{t})dr, \quad \nu = 0,\ldots, \tag{2.9}$$

multiplying the population balance in Eq.2.8 by r^ν, and integrating over all particle sizes, the following system of infinite ordinary differential equations, which describes the rate of change of the moments of the particle size distribution and the solute concentration, is obtained:

$$\begin{aligned}
\frac{d\mu_0}{d\bar{t}} &= -\frac{\mu_0}{\tau} + \left(1 - \frac{4}{3}\pi\mu_3\right) k_2 e^{-\frac{k_3}{\left(\frac{c}{c_s}-1\right)^2}} \\
\frac{d\mu_1}{d\bar{t}} &= -\frac{\mu_1}{\tau} + \nu k_1(c-c_s)\mu_0 \\
\frac{d\mu_2}{d\bar{t}} &= -\frac{\mu_2}{\tau} + \nu k_1(c-c_s)\mu_1 \\
\frac{d\mu_3}{d\bar{t}} &= -\frac{\mu_3}{\tau} + \nu k_1(c-c_s)\mu_2 \\
\frac{d\mu_\nu}{d\bar{t}} &= -\frac{\mu_\nu}{\tau} + \nu k_1(c-c_s)\mu_{\nu-1}, \quad \nu = 4,5,6\ldots, \\
\frac{dc}{d\bar{t}} &= \frac{c_0 - c - 4\pi\tau(c-c_s)\mu_2(\rho - c)}{\tau\left(1 - \frac{4}{3}\pi\mu_3\right)}
\end{aligned} \tag{2.10}$$

Introducing the following set of dimensionless variables and parameters:

$$\begin{gathered}
t = \frac{\bar{t}}{\tau}, \ \tilde{x}_0 = 8\pi\sigma^3\mu_0, \ \tilde{x}_1 = 8\pi\sigma^2\mu_1, \ \tilde{x}_2 = 4\pi\sigma\mu_2, \ \tilde{x}_3 = \frac{4}{3}\pi\mu_3, \ \ldots, \\
\sigma = k_1\tau(c_0 - c_s), \ Da = 8\pi\sigma^3 k_2\tau, \ F = \frac{k_3 c_s^2}{(c_0-c_s)^2}, \ \alpha = \frac{(\rho - c_s)}{(c_0 - c_s)}, \\
\tilde{y} = \frac{(c-c_s)}{(c_0-c_s)}
\end{gathered} \tag{2.11}$$

the following dimensionless system is obtained:

$$\begin{aligned}
\frac{d\tilde{x}_0}{dt} &= -\tilde{x}_0 + (1-\tilde{x}_3)Dae^{\frac{-F}{\tilde{y}^2}} \\
\frac{d\tilde{x}_1}{dt} &= -\tilde{x}_1 + \tilde{y}\tilde{x}_0 \\
\frac{d\tilde{x}_2}{dt} &= -\tilde{x}_2 + \tilde{y}\tilde{x}_1 \\
\frac{d\tilde{x}_3}{dt} &= -\tilde{x}_3 + \tilde{y}\tilde{x}_2 \\
\frac{d\tilde{x}_\nu}{dt} &= -\tilde{x}_\nu + \tilde{y}\tilde{x}_{\nu-1}, \ \nu = 4, \ldots, \\
\frac{d\tilde{y}}{dt} &= \frac{1-\tilde{y}-(\alpha-\tilde{y})\tilde{y}\tilde{x}_2}{1-\tilde{x}_3}
\end{aligned} \tag{2.12}$$

On the basis of the system of Eq.2.12, it is clear that the moments of order four and higher do not affect those of order three and lower, and moreover, the state of the infinite dimensional system:

$$\frac{d\tilde{x}_\nu}{dt} = -\tilde{x}_\nu + \tilde{y}\tilde{x}_{\nu-1}, \ \nu = 4, \ldots, \tag{2.13}$$

is bounded when x_3 and y are bounded, and it converges to a globally exponentially stable equilibrium point when $\lim_{t\to\infty} x_3 = c_1$ and $\lim_{t\to\infty} \tilde{y} = c_2$, where c_1, c_2 are constants. This implies that the dominant dynamics (that is, dynamics associated with eigenvalues that are close to the imaginary axis) of the process of Eq.2.8 can be adequately captured by the following fifth-order moment model:

$$\begin{aligned}
\frac{d\tilde{x}_0}{dt} &= -\tilde{x}_0 + (1-\tilde{x}_3)Dae^{\frac{-F}{\tilde{y}^2}} \\
\frac{d\tilde{x}_1}{dt} &= -\tilde{x}_1 + \tilde{y}\tilde{x}_0 \\
\frac{d\tilde{x}_2}{dt} &= -\tilde{x}_2 + \tilde{y}\tilde{x}_1 \\
\frac{d\tilde{x}_3}{dt} &= -\tilde{x}_3 + \tilde{y}\tilde{x}_2 \\
\frac{d\tilde{y}}{dt} &= \frac{1-\tilde{y}-(\alpha-\tilde{y})\tilde{y}\tilde{x}_2}{1-\tilde{x}_3}
\end{aligned} \tag{2.14}$$

The ability of the above fifth-order moment model to reproduce the dynamics, and to some extent the solutions, of the distributed model of

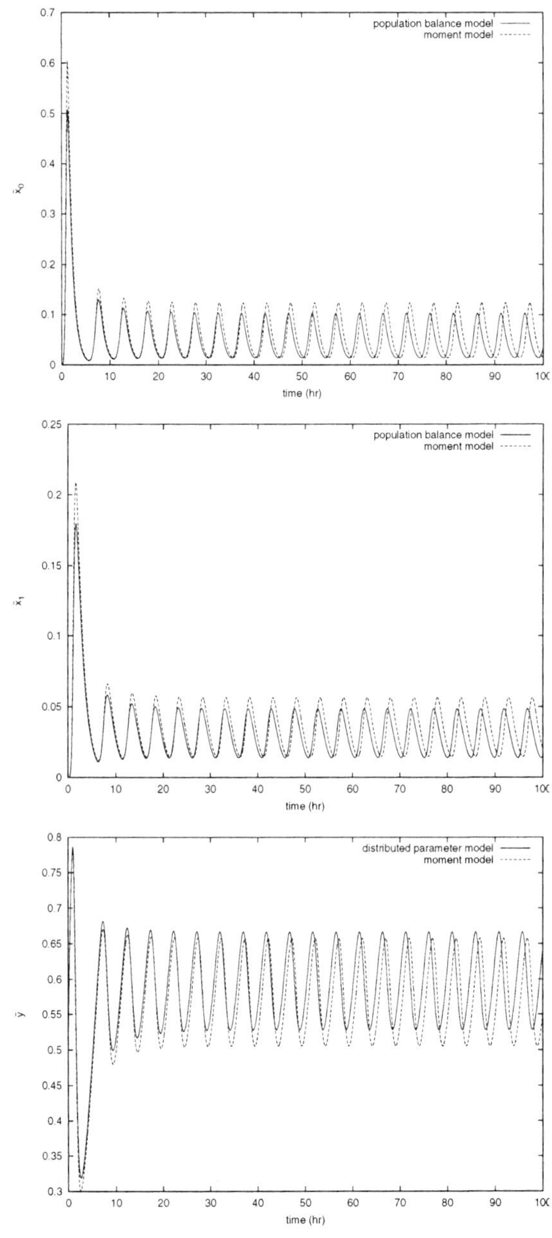

Figure 2.3. Comparison of open-loop profiles of crystal concentration (top), total crystal size (middle) and solute concentration (bottom) obtained from the distributed parameter model and the moment model.

Eq.2.6 is shown in Figure 2.3, where the profiles of the total particle concentration generated by the two models are compared (both models start from the same initial conditions). Even though the discrepancy of the total particle concentration profiles predicted by the two models increases with time (this is expected due to the open-loop instability of the process; see Remark 2.7 below), it is clear that the fifth-order moment model of Eq.2.14 provides a very good approximation of the distributed model of Eq.2.6, thereby establishing that the dominant dynamics of the system of Eq.2.8 are low-dimensional and motivating the use of the moment model for nonlinear controller design (see Chapter 3 for the design and implementation of a nonlinear controller based on the fifth-order model of Eq.2.14).

Remark 2.3: The stability properties of the fifth-order model of Eq.2.14 have been thoroughly studied by Jerauld et al., 1983 (see also Lei et al., 1971), where it has been shown that the global phase space of this model consists of a unique unstable steady-state surrounded by a stable periodic orbit, and that the linearization of the system of Eq.2.6 around the unstable steady-state includes two isolated complex conjugate eigenvalues with a positive real part. For the dimensionless parameters of Table 2.2, the unique unstable steady-state is $\tilde{x}_0 = 0.047$, $\tilde{x}_1 = 0.028$, $\tilde{x}_2 = 0.017$, $\tilde{x}_3 = 0.01$, $\tilde{y} = 0.5996$.

Remark 2.4: We note that even though the number of discretization points, 1,000, used to solve the system of Eq.2.6 is very large (owing to the poor convergence properties of the finite difference scheme), the computation of an accurate (that is, independent of the discretization) solution is critical for the thorough evaluation of the performance of a nonlinear feedback controller synthesized on the basis of a low-order approximation of the distributed parameter system of Eq.2.6 (see Chapter 3 for closed-loop system simulations under nonlinear low-order output feedback control). The adequacy of 1,000 discretization points to yield an accurate solution is established in Figure 2.4, where the profiles of the total crystal concentration and total crystal size for 800 and 1,000 discretization points are compared and are shown to be almost identical.

Remark 2.5: Despite the fact that the model of Eq.2.6 consists of a first-order hyperbolic PDE (population balance) coupled with a nonlinear integro-differential equation (solute mass balance), it is evident from the above dynamical analysis, and the results of the simulation study, that its open-loop dynamic behavior is completely different than the one usually exhibited by systems of first-order hyperbolic PDEs (which arise in the modeling of convection-reaction processes). More specifically, the

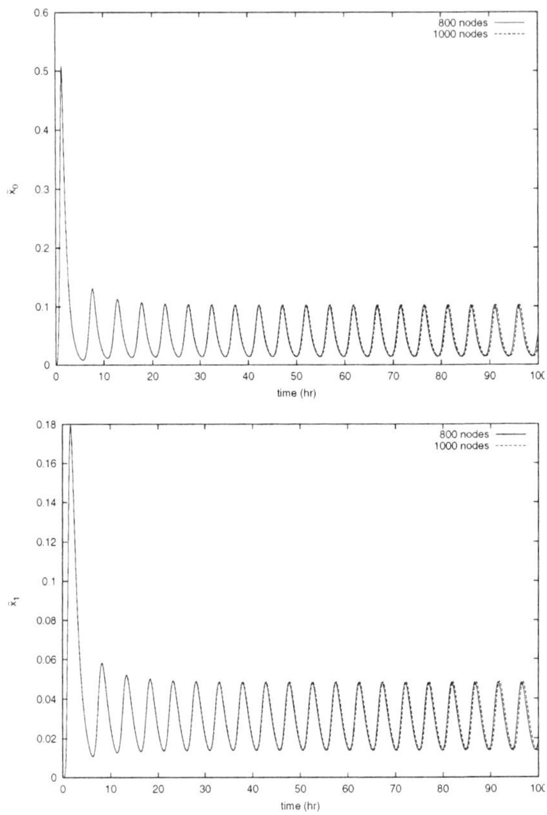

Figure 2.4. Effect of the number of discretization points on crystal concentration (top) and total crystal size (bottom); distributed parameter model.

dominant dynamic behavior of the system of Eq.2.6 is characterized by a small number of degrees of freedom (and thus, it can be described by low-order ODE systems). On the other hand, first-order hyperbolic PDE systems involve spatial differential operators whose eigenvalues cluster along vertical, or nearly vertical, asymptotes in the complex plane, and therefore, they do not exhibit low-dimensional dynamic behavior. This fundamental difference on the nature of the dynamic behavior between first-order hyperbolic PDE systems and particulate process models motivates employing fundamentally different approaches for designing controllers for such systems. In particular, for first-order hyperbolic PDEs, the controller design problem is addressed directly on the basis of the PDE system (see Christofides and Daoutidis, 1996 for results on nonlinear feedback control of hyperbolic PDEs), while for systems of Eqs.2.1-

2.2, the controller design problem will be addressed on the basis of low-order approximations that capture the dominant dynamics, as outlined in section 3.1 in the next section.

Remark 2.6: When a fines trap is used to remove fines, small crystals, from the crystal magma (refer to Lei et al., 1971; Randolph et al., 1987 for a detailed description of crystallizer with fines trap), the model of the crystallizer takes the following form (Lei et al., 1971):

$$\begin{aligned} \frac{\partial n}{\partial \bar{t}} &= -k_1(c-c_s)\frac{\partial n}{\partial r} - \frac{n}{\tau} - \bar{h}(r)\frac{n}{\tilde{\tau}} + \delta(r-0)\bar{\epsilon}k_2 e^{-\frac{k_3}{\left(\frac{c}{c_s}-1\right)^2}} \\ \frac{dc}{d\bar{t}} &= \frac{(c_0-\rho)}{\bar{\epsilon}\tau} + \frac{(\rho-c)}{\tau} + \frac{(\rho-c)}{\bar{\epsilon}}\frac{d\bar{\epsilon}}{d\bar{t}} \end{aligned} \tag{2.15}$$

where $1/\tilde{\tau} = F_0/V$ is the rate at which crystals are circulated through the fines trap (F_0 is the fines recirculation rate and V is the active volume of the crystallizer), and $\bar{h}(r)$ expresses the desired selection curve for fines destruction (classification function). For example, if we desire to remove crystals of size r_m and smaller, then $\bar{h}(r)$ has the form:

$$\bar{h}(r) = \begin{bmatrix} 1, & for \;\; r \leq r_m \\ 0, & for \;\; r > r_m \end{bmatrix} \tag{2.16}$$

For the distributed parameter model of Eq.2.15 with the above expression for $\bar{h}(r)$, one can easily show that a direct application of the method of moments leads to an unclosed set of moments equations, thereby implying the need of using a general model reduction procedure for particulate processes (see section 3.2 and the crystallizer application in Chapter 4).

3. Nonlinear Model Reduction and Control of Particulate Processes

3.1 Methodological Framework

Owing to its distributed parameter nature, the system of Eqs.2.1-2.2 cannot be directly used as the basis for the synthesis of low-order nonlinear controllers that can be implemented. This fact, together with the realization that the dominant dynamics of particulate processes are characterized by a small number of degrees of freedom, motivates employing the following methodology for the synthesis of low-order nonlinear output feedback controllers for systems of the form of Eqs.2.1-2.2:

1 Initially, the method of weighted residuals is used to derive a nonlinear, possibly high-order, ODE system that accurately reproduces the solutions and dynamics of the system of Eqs.2.1-2.2. Then, a procedure based on the concept of approximate inertial manifold is employed for the construction of low-order ODE systems that accurately reproduce the dominant dynamics of the large-scale ODE system obtained by the method of weighted residuals. The asymptotic validity of the ODE approximations is established by using results from the perturbation theory.

2 Then, the low-order ODE approximation of the system of Eqs.2.1-2.2 is used as the basis for the synthesis, via geometric control methods, of nonlinear output feedback controllers that stabilize the closed-loop ODE system and enforce set-point tracking.

3 Finally, the resulting closed-loop system (particulate process model of Eqs.2.1-2.2 and controller) is analyzed to derive conditions that guarantee that the desired stability and set-point tracking properties are enforced in the infinite dimensional closed-loop system.

3.2 Nonlinear Model Reduction of Particulate Process Models

In this section, we introduce a general methodology for deriving low-order ODE systems that accurately reproduce the dominant dynamics of the nonlinear integro-differential equation system of Eqs.2.1-2.2. The proposed model reduction methodology exploits the low-dimensional behavior of the dominant dynamics of the system of Eqs.2.1-2.2 and is based on a combination of the method of weighted residuals with the concept of approximate inertial manifold.

3.2.1 Method of Weighted Residuals

We initially use the method of weighted residuals to construct a nonlinear, possibly high-order, ODE system that accurately reproduces the solutions and dynamics of the distributed parameter system of Eqs.2.1-2.2. The central idea of the method of weighted residuals (see Ramkrishna, 1985 for a comprehensive review of results on the use of this method for solving population balance equations) is to approximate the exact solution of $n(r,t)$ by an infinite series of orthogonal basis functions defined on the interval $[0, r_{max})$ with time-varying coefficients, substitute the series expansion into Eq.2.1 to form the residual, and then force the residual to be orthogonal to a complete set of weighted functions (that is, the inner product of the residual with a complete set of weighting functions in $L_2[0, r_{max})$ is set equal to zero) to compute a set of ODEs which describe the rate of change of the time-varying coefficients of the series expansion of the solution.

Specifically, we consider an orthogonal set of basis functions $\phi_k(r)$, where $r \in [0, r_{max})$, $k = 1, \dots, \infty$, and expand the particle size distribution function $n(r,t)$ in an infinite series in terms of $\phi_k(r)$ as follows:

$$n(r,t) = \sum_{k=1}^{\infty} a_k(t)\phi_k(r) \tag{2.17}$$

where $a_k(t)$ are time-varying coefficients. Substituting Eq.2.17 into Eqs.2.1-2.2, we get:

$$\begin{aligned}\sum_{k=1}^{\infty}\phi_k(r)\frac{\partial a_k(t)}{\partial t} &= -\sum_{k=1}^{\infty}a_k(t)\frac{\partial(G(x,r)\phi_k(r))}{\partial r} + w(\sum_{k=1}^{\infty}a_k(t)\phi_k(r),x,r)\\ \dot{x} &= f(x) + g(x)u(t) + A\int_0^{r_{max}} a(\sum_{k=1}^{\infty}a_k(t)\phi_k(r),r,x)dr\end{aligned} \tag{2.18}$$

Multiplying the population balance with the weighting functions $\psi_\nu(r)$, and integrating over the entire particle size spectrum (that is, taking inner product in $L_2[0, r_{max})$ with the weighting functions), the following set of infinite ODEs is obtained:

$$\begin{aligned}&\int_0^{r_{max}}\psi_\nu(r)\sum_{k=1}^{\infty}\phi_k(r)\frac{\partial a_k(t)}{\partial t}dr =\\ &\quad -\sum_{k=1}^{\infty}a_k(t)\int_0^{r_{max}}\psi_\nu(r)\frac{\partial(G(x,r)\phi_k(r))}{\partial r}dr\\ &\quad +\int_0^{r_{max}}\psi_\nu(r)w(\sum_{k=1}^{\infty}a_k(t)\phi_k(r),x,r)dr,\ \nu = 1,\dots,\infty\\ &\dot{x} = f(x) + g(x)u(t) + A\int_0^{r_{max}} a(\sum_{k=1}^{\infty}a_k(t)\phi_k(r),r,x)dr\end{aligned} \tag{2.19}$$

Truncating the series expansion of $n(r,t)$ up to order N and keeping the first N equations (that is, $\nu = 1, \dots, N$), the infinite dimensional system of Eq.2.19 reduces to the following finite set of ODEs:

$$
\begin{aligned}
&\int_0^{r_{max}} \psi_\nu(r) \sum_{k=1}^{N} \phi_k(r) \frac{\partial a_{kN}(t)}{\partial t} dr = \\
&\qquad -\sum_{k=1}^{N} a_{kN}(t) \int_0^{r_{max}} \psi_\nu(r) \frac{\partial (G(x_N, r)\phi_k(r))}{\partial r} dr \\
&\qquad + \int_0^{r_{max}} \psi_\nu(r) w(\sum_{k=1}^{N} a_{kN}(t)\phi_k(r), x_N, r) dr, \ \nu = 1, \ldots, N \\
&\dot{x}_N = f(x_N) + g(x_N)u(t) + A \int_0^{r_{max}} a(\sum_{k=1}^{N} a_{kN}(t)\phi_k(r), r, x_N) dr \\
&y_i(t) = h_i(\int_0^{r_{max}} c_\kappa \sum_{k=1}^{N} a_{kN}(t)\phi_k(r) dr, x_N), \\
&\quad i = 1, \ldots, m, \ \ \kappa = 1, \ldots, l
\end{aligned}
\tag{2.20}
$$

where x_N and a_{kN} are the approximations of x and a_k obtained by an N-th order truncation. From Eq.2.20, it is clear that the form of the ODEs that describe the rate of change of $a_{kN}(t)$ depends on the choice of the basis and weighting functions, as well as on N. The basis and weighting functions determine the type of weighted residual method being used (see Remarks 2.8 and 2.9 below).

Proposition 2.1 that follows establishes a convergence property for the discrepancy between the solutions of the particulate process model of Eqs.2.1-2.2 and the approximation of Eq.2.20, for sufficiently large N. The proof is given in Appendix 9.A.

Proposition 2.1: *Consider the system of Eqs.2.1-2.2 with $u(t) \equiv 0$ and assume that n and $\frac{\partial n}{\partial r}$ are continuous functions of r. Suppose also that the system of Eq.2.20 is locally exponentially stable, for any N. Then, there exists an N sufficiently large so that $\forall t \geq 0$:*

$$
\begin{aligned}
n(r,t) &= n_N(r,t) + O(\epsilon(N)) \\
x(t) &= x_N(t) + O(\epsilon(N))
\end{aligned}
\tag{2.21}
$$

where $\epsilon(N)$ is a small positive real number that depends on N and satisfies $\lim_{N\to\infty} \epsilon(N) = 0$, and $n_N(r,t) = \sum_{k=1}^{N} a_{kN}(t)\phi_k(r)$ is the approximation of $n(r,t)$ which is obtained by solving Eq.2.20 with $u(t) \equiv 0$.

Remark 2.7: The assumption that the system of Eq.2.20 (and thus, the system of Eqs.2.1-2.2) is locally exponentially stable is necessary in order to prove that the estimates of Eq.2.21 hold for all times. When the system of Eq.2.20 is not exponentially stable, one can only prove that the estimates of Eq.2.21 hold for $t \in [0, \tau]$ where τ is a positive real number of $O(1)$ (see simulation results in section 2.2 for a verification of this fact). Furthermore, the convergence result can be shown for the $L_2[0, \infty)$ norm of the difference between the solution of the population balance $n(r, t)$ and its approximation of Eq.2.17, that is, for all $t \in [0, \infty)$:

$$\lim_{N \to \infty} ||n(r,t) - \sum_{k=1}^{N} a_{kN}(t)\phi_k(r)||_2 = 0 \tag{2.22}$$

Remark 2.8: The method of weighted residuals reduces to the method of moments when the basis functions are chosen to be Laguerre polynomials and the weighting functions are chosen as $\psi_\nu = r^\nu$. The moments of the particle size distribution are defined as:

$$\mu_\nu = \int_0^\infty r^\nu n(r,t)dr, \quad \nu = 0, \ldots, \infty \tag{2.23}$$

and the moment equations can be directly generated from the population balance model by multiplying it by r^ν, $\nu = 0, \ldots, \infty$ and integrating from 0 to ∞. The procedure of forming moments of the population balance equation very often leads to terms that may not reduce to moments, terms that include fractional moments, or to an unclosed set of moment equations. To overcome this problem, the particle size distribution is expanded in terms of Laguerre polynomials defined in $L_2[0, \infty)$ and the series solution is used to close the set of moment equations (this procedure can be used for models of crystallizers with fines trap; see Eqs.2.15-2.16).

Remark 2.9: When the number of basis functions $\phi_k(r)$ required to obtain a good approximation (measured in a desired norm) of the solution of the population balance, is small, then the weighting functions are usually chosen (Rawlings et al., 1993) to be identical to the basis functions, in which case the method of weighted residuals reduces to Galerkin's method.

Remark 2.10: When a "good" (in the sense of leading to the derivation of a moderate order ODE system of desired accuracy) set of basis functions $\phi_k(r)$ cannot be found within the standard basis function sets, one can compute a set of empirical eigenfunctions by applying the Karhunen-Loeve expansion (also known as proper orthogonal decomposition; see

Fukunaga, 1990; Holmes et al., 1996; Shvartsman and Kevrekidis, 1998, for details) on an appropriate ensemble of solutions of the particulate process model of Eqs.2.1-2.2, which are obtained from detailed finite difference discretizations.

3.2.2 Inertial Manifold / Approximate Inertial Manifold

The system of Eq.2.20 was obtained from a direct application of the method of weighted residuals (with arbitrary basis functions) to the system of Eqs.2.1-2.2, and thus, may be of very high order in order to provide an accurate description of the dominant dynamics of the particulate process model. High-dimensionality of the system of Eq.2.20 leads to complex controller design and high-order controllers, which cannot be readily implemented in practice. To circumvent these problems, we exploit the low-dimensional behavior of the dominant dynamics of particulate processes and use an approach based on the concept of inertial manifold to derive low-order ODE systems that accurately describe the dominant dynamics of the system of Eq.2.20. The concept of inertial manifold is an appropriate tool for model reduction because if the trajectories of the system of Eq.2.20 are on the manifold, then this system is *exactly* described by a low-order system.

We begin with the definition of the concept of inertial manifold, and we continue with the concept of approximate inertial manifold. To this end, we exploit the orthogonality of the basis functions $\phi_k(r)$ to uniquely split:

$$n_N = \sum_{k=1}^{p} a_{kN}(t)\phi_k(r) + \sum_{k=p+1}^{N} a_{kN}(t)\phi_k(r) =: n_p + n_q \qquad (2.24)$$

where the first p eigenmodes are associated with the dominant (possibly unstable) dynamics of the system of Eq.2.20, and the remaining q eigenmodes are associated with exponentially stable dynamics.

Introducing the vector notation $a_N = [a_{1N} \ \cdots \ a_{NN}]$ and defining the new vectors of state variables $\tilde{x} = [a_{sN}^T \ x_N^T]^T$, $a_{sN} = [a_{1N} \ \cdots a_{pN}]^T$ and $a_{fN} = [a_{(p+1)N} \ \cdots a_{NN}]^T$, the system of Eq.2.20 can be written as:

$$\begin{aligned} \dot{a}_{fN} &= \bar{f}_q(a_{fN}, \tilde{x}) \\ \dot{\tilde{x}} &= \tilde{f}(\tilde{x}, a_{fN}) + \tilde{g}(\tilde{x}, a_{fN})u \\ y_i &= \tilde{h}_i(\tilde{x}, a_{fN}), \quad i = 1, \ldots, m \end{aligned} \qquad (2.25)$$

where the explicit expression of $\bar{f}_q(a_{fN}, \tilde{x}), \tilde{f}(\tilde{x}, a_{fN}), \tilde{g}(\tilde{x}, a_{fN}), \tilde{h}_i(\tilde{x}, a_{fN})$ can be obtained by comparing Eq.2.20 and Eq.2.25 and will be omitted for brevity. Assuming that $\bar{f}_q(a_{fN}, \tilde{x}) = \tilde{A}a_{fN} + \tilde{f}_q(a_{fN}, \tilde{x})$, where $\tilde{A}$

is a Hurwitz matrix and $\tilde{f}_q(a_{fN}, \tilde{x})$ is a nonlinear vector function which does not include linear terms (this assumption is made to simplify the development and can be readily relaxed), the system of Eq.2.25 can be written as:

$$\begin{aligned} \dot{a}_{fN} &= \tilde{A}a_{fN} + \tilde{f}_q(a_{fN}, \tilde{x}) \\ \dot{\tilde{x}} &= \tilde{f}(\tilde{x}, a_{fN}) + \tilde{g}(\tilde{x}, a_{fN})u \\ y_i &= \tilde{h}_i(\tilde{x}, a_{fN}), \quad i = 1, \ldots, m \end{aligned} \tag{2.26}$$

For the above system, an inertial manifold $\mathcal{M}$ is a subset of $\mathbb{R}^{N+n}$, which satisfies the following properties (Temam, 1988):

i) $\mathcal{M}$ is a finite dimensional Lipschitz manifold,

ii) $\mathcal{M}$ is a graph of a Lipschitz function $\Sigma(\tilde{x})$ mapping $\mathbb{R}^{p+n}$ into $\mathbb{R}^q$ and for every solution $\tilde{x}(t), a_{fN}(t)$ of Eq.2.26 with $a_{fN}(0) = \Sigma(\tilde{x}(0))$, then

$$a_{fN}(t) = \Sigma(\tilde{x}(t)), \quad \forall\, t \geq 0 \tag{2.27}$$

iii) $\mathcal{M}$ attracts every trajectory exponentially.

The evolution of the state $a_{fN}(t)$ on $\mathcal{M}$ is given by Eq.2.27, while the evolution of the state $\tilde{x}$ is governed by the following $(p+n)$-order system:

$$\begin{aligned} \dot{\tilde{x}} &= \tilde{f}(\tilde{x}, \Sigma(\tilde{x})) + \tilde{g}(\tilde{x}, \Sigma(\tilde{x}))u \\ y_i &= \tilde{h}_i(\tilde{x}, \Sigma(\tilde{x})), \quad i = 1, \ldots, m \end{aligned} \tag{2.28}$$

Differentiating Eq.2.27 and utilizing Eq.2.26, $\Sigma(\tilde{x})$ can be computed as the solution of the following partial differential equation:

$$\frac{\partial \Sigma}{\partial \tilde{x}}[\tilde{f}(\tilde{x}, \Sigma(\tilde{x})) + \tilde{g}(\tilde{x}, \Sigma(\tilde{x}))u] = \tilde{A}\Sigma(\tilde{x}) + \tilde{f}_q(\Sigma(\tilde{x}), \tilde{x}) \tag{2.29}$$

which $\Sigma(\tilde{x})$ has to satisfy for all $\tilde{x} \in \mathbb{R}^{p+n}$. From the structure of Eq.2.29, it is obvious that the computation of the explicit form of $\Sigma(\tilde{x})$ is a very difficult (if not impossible) task in most practical applications. To overcome this problem, and since the dynamics of the a_{fN} modes are stable and faster than the ones of the $\tilde{x}$ modes (note that the number of unstable eigenvalues of the system of Eqs.2.1-2.2 is finite, and by assumption, all the unstable eigenvalues are included in the $\tilde{x}$ subsystem), we obtain $\Sigma(\tilde{x})$ by setting $\dot{a}_{fN} = \frac{\partial \Sigma}{\partial \tilde{x}}[\tilde{f}(\tilde{x}, \Sigma) + \tilde{g}(\tilde{x}, \Sigma)u] \equiv 0$ and solving the equation:

$$\tilde{A}\Sigma(\tilde{x}) + \tilde{f}_q(\Sigma(\tilde{x}), \tilde{x}) = 0 \tag{2.30}$$

using a standard successive approximation (fixed point) algorithm of Foias and Témam, 1988 (see also Shvartsman and Kevrekidis, 1998):

$$\begin{aligned} \tilde{\Sigma}_{\kappa+1}(\tilde{x}) &= -\tilde{A}^{-1}\tilde{f}_q(\tilde{\Sigma}_{\kappa}(\tilde{x}),\tilde{x}), \quad \kappa=0,\ldots,\bar{l}, \quad \tilde{\Sigma}_0(\tilde{x})=0 \\ \tilde{a}_{fN} &= \tilde{\Sigma}_{\bar{l}+1}(\tilde{x}) \end{aligned} \tag{2.31}$$

where $\tilde{\Sigma}_{\bar{l}+1}(\tilde{x})$ is an approximation of $\Sigma(\tilde{x})$ (called approximate inertial manifold) and $\tilde{a}_{fN}$ is the approximation of a_{fN}. Substituting $\tilde{\Sigma}_{\bar{l}+1}(\tilde{x})$ into the system of Eq.2.28, the following $(p+n)$-order approximation of the particulate process model is obtained:

$$\begin{aligned} \dot{\tilde{x}} &= \tilde{f}(\tilde{x},\tilde{\Sigma}_{\bar{l}+1}(\tilde{x})) + \tilde{g}(\tilde{x},\tilde{\Sigma}_{\bar{l}+1}(\tilde{x}))u =: \tilde{f}(\tilde{x}) + \tilde{g}(\tilde{x})u \\ y_{si} &= \tilde{h}_i(\tilde{x},\tilde{\Sigma}_{\bar{l}+1}(\tilde{x})) =: \tilde{h}_i(\tilde{x}), \quad i=1,\ldots,m \end{aligned} \tag{2.32}$$

where the subscript s in the controlled output, y_{si}, denotes that this controlled output is associated with an approximate low-order ODE system (refer to Eq.2.3 for the formulation of y_i for the distributed parameter model).

Proposition 2.2 that follows establishes that the local stability properties and solutions, for large times, of the systems Eq.2.32 and Eq.2.20 are equivalent. The proof is given in appendix 9.A.

Proposition 2.2: *Suppose that the sequence of Eq.2.31, for the construction of $\tilde{\Sigma}_{\bar{l}+1}(\tilde{x})$, converges for $\bar{l}$ sufficiently large (that is, for any $\tilde{\epsilon}$, there exists an $\bar{l}^*$ such that if $\bar{l} \geq \bar{l}^*$, then $|\Sigma(\tilde{x}) - \tilde{\Sigma}_{\bar{l}+1}(\tilde{x})| \leq \tilde{\epsilon}$). Suppose also that the $(p+n)$-order system of Eq.2.32 with $u(t) \equiv 0$ is locally exponentially stable. Then, the $(N+n)$-order system of Eq.2.20 with $u(t) \equiv 0$ is locally exponentially stable and $\lim_{t\to\infty} ||n_N - \bar{n}_N||_2 = O(\tilde{\epsilon}(\bar{l}))$ where $\bar{n}_N = \sum_{k=1}^{p} \tilde{a}_{kN}(t)\phi_k(r) + \sum_{k=p+1}^{N} \tilde{\Sigma}_{\bar{l}+1}(\tilde{x})\phi_k(r)$ ($\tilde{a}_{kN}(t)$ is the solution obtained from the system of Eq.2.32) and $\tilde{\epsilon}(\bar{l})$ is a small positive real number that depends on $\bar{l}$ and satisfies $\lim_{\bar{l}\to\infty} \tilde{\epsilon}(\bar{l}) = 0$.*

Remark 2.11: We note that even though many particulate processes exhibit low-dimensional dynamic behavior, the delicate mathematical question of rigorously establishing existence of inertial manifolds for particulate process models, at this stage, is unresolved. Such a question has been positively answered for certain classes of diffusion-reaction systems and the Kuramoto-Sivashinsky equation (see Temam, 1988 for details).

Remark 2.12: The expression of the approximate inertial manifold $\tilde{\Sigma}_{\bar{l}+1}(\tilde{x})$ of Eq.2.31 (where $\bar{l}$ is chosen based on the desired degree of approximation) was originally proposed in Foias and Témam, 1988 and is called the steady manifold. Also refer to Christofides and Daoutidis, 1997; Shvartsman and Kevrekidis, 1998; Christofides, 2001 for alternative expressions of $\tilde{\Sigma}(\tilde{x})$, as well as detailed computational studies that show that the use of approximate inertial manifolds leads to accurate low-order ODE approximations and low-order controllers for diffusion-reaction systems described by parabolic partial differential equations.

Remark 2.13: For $\Sigma(\tilde{x}) = 0$, we obtain the following order $(p+n)$ approximation of the particulate process model:

$$\begin{aligned} \dot{a}_{sN} &= \bar{f}_p(a_{sN}, 0, x_N) \\ \dot{x}_N &= f(x_N) + g(x_N)u(t) + A\int_0^{r_{max}} a(n_p, x_N, r)dr \end{aligned} \tag{2.33}$$

The above system is identical to the one obtained by a direct application of the method of weighted residuals to the particulate process model of Eqs.2.1-2.2 (Eq.2.20 with $N = p$).

3.3 Nonlinear Output Feedback Control of Particulate Processes

In this section, the system of Eq.2.32 is used to synthesize a nonlinear finite dimensional output feedback controller that guarantees stability and enforces output tracking in the closed-loop ODE system and to establish that the same controller exponentially stabilizes the closed-loop particulate process model. The output feedback controller is constructed through a standard combination of a state feedback controller with a state observer. The state feedback controller is synthesized via geometric control methods and the state observer is an extended Luenberger-type observer. This begins with some preliminaries, which will be used to state the controller synthesis result.

3.3.1 Preliminaries

Referring to the system of Eq.2.32, we define the relative order of the output y_{si} with respect to the vector of manipulated inputs u as the smallest integer r_i for which

$$\left[L_{\tilde{g}_1} L_{\tilde{f}}^{r_i-1}\tilde{h}_i(\tilde{x}) \ \cdots \ L_{\tilde{g}_m} L_{\tilde{f}}^{r_i-1}\tilde{h}_i(\tilde{x})\right] \not\equiv [0 \ \cdots \ 0] \tag{2.34}$$

where $\tilde{g}_i$ is the i-th vector of the matrix $\tilde{g}$, or $r_i = \infty$ if such an integer does not exist. We also define the characteristic matrix

$$C(\tilde{x}) = \begin{bmatrix} L_{\tilde{g}_1} L_{\tilde{f}}^{r_1-1} \tilde{h}_1(\tilde{x}) & \cdots & L_{\tilde{g}_m} L_{\tilde{f}}^{r_1-1} \tilde{h}_1(\tilde{x}) \\ L_{\tilde{g}_1} L_{\tilde{f}}^{r_2-1} \tilde{h}_2(\tilde{x}) & \cdots & L_{\tilde{g}_m} L_{\tilde{f}}^{r_2-1} \tilde{h}_2(\tilde{x}) \\ \vdots & \cdots & \vdots \\ L_{\tilde{g}_1} L_{\tilde{f}}^{r_m-1} \tilde{h}_m(\tilde{x}) & \cdots & L_{\tilde{g}_m} L_{\tilde{f}}^{r_m-1} \tilde{h}_m(\tilde{x}) \end{bmatrix} \quad (2.35)$$

3.3.2 Controller Synthesis

We use the nonlinear system of Eq.2.32 as a basis for the synthesis, via geometric control methods, of nonlinear state feedback controllers of the general form:

$$u = p(\tilde{x}) + Q(\tilde{x})v \quad (2.36)$$

where $p(\tilde{x})$ is a smooth vector function, $Q(\tilde{x})$ is a smooth matrix, and $v \in \mathbb{R}^m$ is the constant reference input vector. The controllers guarantee local exponential stability and enforce a linear input/output response in the system of Eq.2.32 (the details on controller synthesis can be found in Isidori, 1995 and will be omitted here for brevity).

Under the hypothesis that the system of Eq.2.32 is locally observable (that is, its linearization around the desired operating steady-state is observable), the practical implementation of a nonlinear state feedback controller of the form of Eq.2.36 will be achieved by employing the following nonlinear state observer:

$$\frac{d\omega}{dt} = \tilde{f}(\omega) + \tilde{g}(\omega)u + L(y - \tilde{h}(\omega)) \quad (2.37)$$

where ω denotes the observer state vector (the dimension of the vector ω is equal to the dimension of $\tilde{x}$ in the system of Eq.2.32), $y = [y_1 \ y_2 \ \cdots y_l]^T$ is the measured output vector, and L is a matrix chosen so that the eigenvalues of the matrix $C_L = \frac{\partial \tilde{f}}{\partial \omega}_{(\omega=\omega_s)} - L\frac{\partial \tilde{h}}{\partial \omega}_{(\omega=\omega_s)}$, where ω_s is the operating steady-state, lie in the open left-half of the complex plane. The state observer of Eq.2.37 consists of a replica of the system of Eq.2.32 plus a linear gain multiplying the discrepancy between the actual and the estimated value of the output, and therefore, it is an extended Luenberger-type observer.

The state feedback control law of Eq.2.36 and the state observer of Eq.2.37 can be combined to yield the following nonlinear output feedback control law:

$$\begin{aligned} \frac{d\omega}{dt} &= \tilde{f}(\omega) + \tilde{g}(\omega)(p(\omega) + Q(\omega)v) + L(y - \tilde{h}(\omega)) \\ u &= p(\omega) + Q(\omega)v \end{aligned} \tag{2.38}$$

Theorem 2.1 below provides an explicit synthesis formula of the above output feedback control law and conditions that guarantee closed-loop stability and asymptotic output tracking (the proof can be found in Appendix 9.A).

Theorem 2.1: *Suppose that the sequence of Eq.2.31 converges for $\bar{l}$ is sufficiently large. Consider the system of Eq.2.32 and assume that:* 1) *it is locally observable in the sense that there exists a matrix L such that $C_L = \frac{1}{\mu}\bar{A}$ where μ is a small positive parameter and $\bar{A}$ is a Hurwitz matrix,* 2) *its characteristic matrix, $C(\tilde{x})$, is nonsingular $\forall\ \tilde{x} \in D \subset \mathbb{R}^{p+n}$, and* 3) *its unforced ($v \equiv 0$) zero dynamics are locally exponentially stable. Finally, consider the particulate process model of Eqs.2.1-2.2 under the nonlinear output feedback controller:*

$$\begin{aligned} \frac{d\omega}{dt} &= \tilde{f}(\omega) + \tilde{g}(\omega)\left\{[\beta_{1r_1}\cdots\beta_{mr_m}]C(\omega)\right\}^{-1}\left\{v - \sum_{i=1}^{m}\sum_{k=0}^{r_i}\beta_{ik}L_{\tilde{f}}^{k}\tilde{h}_i(\omega)\right\} \\ &\quad + L(y - \tilde{h}(\omega)) \\ u &= \left\{[\beta_{1r_1}\cdots\beta_{mr_m}]C(\omega)\right\}^{-1}\left\{v - \sum_{i=1}^{m}\sum_{k=0}^{r_i}\beta_{ik}L_{\tilde{f}}^{k}\tilde{h}_i(\omega)\right\} \end{aligned} \tag{2.39}$$

where the parameters β_{ik} are chosen so that the roots of the equation $det(B(s)) = 0$ are in the open left-half of the complex plane ($B(s)$ is an $l \times l$ matrix, whose (i,j)-th element is of the form $\sum_{k=0}^{r_i}\beta_{jk}^{i}s^k$). Then, there exists a positive real number μ^* such that if $\mu \in (0, \mu^*]$, the closed-loop system (particulate process model and controller of Eq.2.39) is exponentially stable and $\lim_{t\to\infty}|y_i - v_i| = O(\hat{\epsilon}(N+\bar{l}))$, where v_i is the set-point for the ith controlled output and $\hat{\epsilon}(N+\bar{l})$ is a small positive real number that depends on $N, \bar{l}$ and satisfies $\lim_{N,\bar{l}\to\infty}\hat{\epsilon}(N+\bar{l}) = 0$.

Remark 2.14: Regarding the practical application of Theorem 2.1, one has to initially pick an $\bar{l}$ and construct the system of Eq.2.32, and then verify assumptions 1, 2, and 3 of the theorem on the basis of this system. If these assumptions are satisfied, the synthesis formula of Eq.2.39 can be directly used to derive the explicit form of the controller (see Chapter 3 for an application of this procedure to the continuous crystallizer example and to a batch crystallizer example).

Remark 2.15: The assumption that $\bar{l}$ is sufficiently large is needed to obtain the stability and closeness of solutions results of Proposition 2.2 for the appropriate approximations of the closed-loop system, while the assumption $C_L = \frac{1}{\mu}\bar{A}$, where μ is a small positive parameter and $\bar{A}$ is a Hurwitz matrix, is needed to ensure that the presence of closed-loop system states, which are not included in the model used for controller synthesis, in the state observer does not lead to closed-loop instability. Finally, the assumption that the characteristic matrix $C(\tilde{x})$ is nonsingular is made to simplify the presentation of the controller synthesis results and can be relaxed (see Isidori, 1995 for details).

Remark 2.16: The exponential stability of the closed-loop system guarantees that in the presence of small initialization errors of the observer states (that is, $\omega(0) \neq \tilde{x}(0)$), and uncertainty in the process parameters and disturbances, the states of the closed-loop system will be bounded. Furthermore, since the number of manipulated inputs and controlled outputs is finite, it is possible to implement a linear error feedback controller (for example, a proportional-integral (PI) controller for single-input single-output processes) around the $(y_i - v_i)$, $i = 1, \ldots, l$, loops to ensure asymptotic offsetless output tracking in the closed-loop system, in the presence of such uncertainty.

Remark 2.17: Theorem 2.1 establishes that a nonlinear output feedback controller which guarantees local exponential stability and output tracking in the finite dimensional closed-loop system (Eqs.2.32-2.39), continues to enforce the same properties in the infinite dimensional closed-loop system (Eqs.2.1-2.2-2.39). This result is intuitively expected because: a) the dynamics of the modes of the particulate process model which are not taken into account in the controller design (that is, not included in the ODE model of Eq.2.32), are locally exponentially stable, and b) the control action $u(t)$ does not influence the dynamics of the modes which are not taken into account in the controller design (note that $u(t)$ does not enter in the a_{fN}-subsystem of Eq.2.25).

Remark 2.18: The nonlinear controller of Eq.2.39 possesses a robustness property with respect to fast and asymptotically stable unmodeled dynamics (that is, the controller enforces exponential stability and output tracking in the closed-loop system despite the presence of additional dynamics in the process, as long as they are stable and sufficiently fast). This property of the controller of Eq.2.39 can be rigorously established by analyzing the closed-loop system with the unmodeled dynamics using singular perturbations and is of particular importance for many practical applications where unmodeled dynamics often occur due to actuator and sensor dynamics, fast process dynamics, and so on.

4. Conclusions

In this chapter, we considered spatially-homogeneous particulate processes modeled by a class of nonlinear partial integro-differential equation systems and developed a general method for the synthesis of practically implementable nonlinear output feedback controllers. The controllers enforce stability and attain a particle size distribution with desired characteristics in the closed-loop system. Initially, a model reduction procedure based on a combination of the method of weighted residuals and the concept of approximate inertial manifold was presented for the construction of low-order ODE systems that accurately reproduce the dynamics of the particulate process. These ODE systems were then used for the synthesis of nonlinear low-order output feedback controllers that enforce exponential stability in the closed-loop system and achieve a desired particle size distribution. In the next chapter, the proposed control method is applied to continuous and batch crystallization systems and its performance and robustness properties are tested through simulations.

Chapter 3

NONLINEAR CONTROL OF CRYSTALLIZATION

In this chapter, we present two applications of the nonlinear control method proposed in Chapter 2 to continuous and batch crystallization systems. Low-dimensional approximations of the population balances are derived and used for nonlinear controller design. The performance of the nonlinear controllers is successfully tested through simulations using the detailed population balance models and is shown to be superior to the one of linear control schemes.

1. Application to a Continuous Crystallizer

1.1 Nonlinear Controller Synthesis

In this section, the proposed nonlinear control method is used to stabilize the continuous crystallizer introduced in section 2.2. Motivated by the fact that the crystallizer with the crystal size distribution as controlled output and the solute feed concentration as manipulated input is an approximately controllable system (see Semino and Ray, 1995a for a rigorous controllability analysis), we study two representative control problems: the first one involves manipulating the solute feed concentration, $u(t) = c_0 - c_{0s}$ where c_{0s} is the steady-state solute feed concentration, to achieve a crystal size distribution with desired mass, that is, the controlled output is defined as:

$$y(\bar{t}) \quad = \quad 8\pi\sigma^3 \int_0^{\infty} n(r,\bar{t})dr = \tilde{x}_0 \tag{3.1}$$

and the second one involves manipulating the solute feed concentration to achieve a crystal size distribution with a desired total particle size, that is, the controlled output is defined as:

$$y(\bar{t}) \;=\; 8\pi\sigma^2 \int_0^{\infty} rn(r,\bar{t})dr = \tilde{x}_1 \tag{3.2}$$

Refer to Lei et al., 1971; Randolph et al., 1987; Eaton and Rawlings, 1990; Rawlings et al., 1993 for the use of other manipulated variables including fines destruction rate and crystallizer temperature for the stabilization of crystallizers (note that the proposed control method can be used for the synthesis of nonlinear controllers when such manipulated inputs are considered).

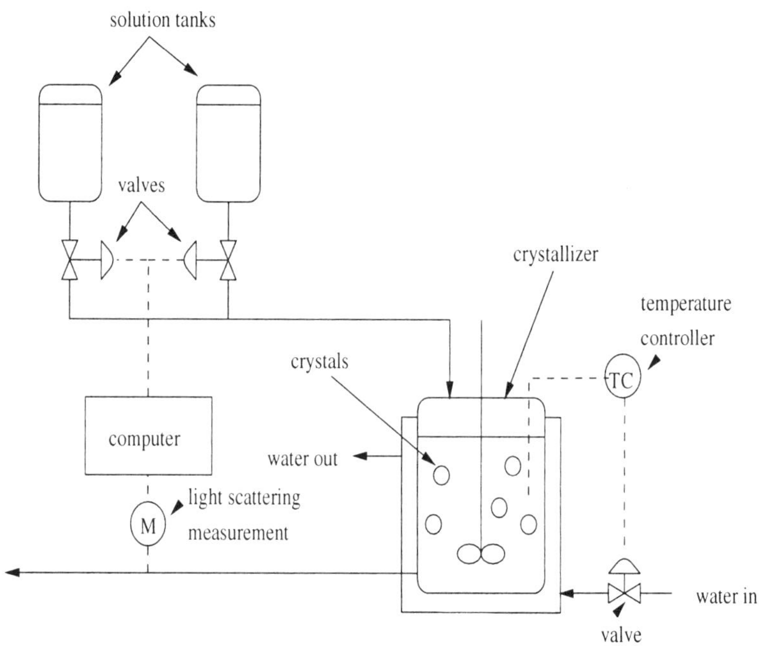

Figure 3.1. Schematic diagram for the proposed process.

Including the above specifications for manipulated input and controlled outputs in the model of Eq.2.8, one can easily see that the resulting system is in the form of Eq.2.5. Applying the method of moments to this system, utilizing the dimensionless variables of Eq.2.11 and $\bar{u} = \dfrac{(c_0 - c_{0s})}{(c_0 - c_s)}$ and neglecting the moments of order four and higher, one can derive a system of the form of Eq.2.25 with $\tilde{x} = [\tilde{x}_0 \; \tilde{x}_1 \; \tilde{x}_2 \; \tilde{x}_3 \; \tilde{y}]^T$

and:

$$\tilde{f}(\tilde{x}) = \begin{bmatrix} -\tilde{x}_0 + (1-\tilde{x}_3)Dae^{\frac{-F}{\tilde{y}^2}} \\ -\tilde{x}_1 + \tilde{y}\tilde{x}_0 \\ -\tilde{x}_2 + \tilde{y}\tilde{x}_1 \\ -\tilde{x}_3 + \tilde{y}\tilde{x}_2 \\ \dfrac{1-\tilde{y}-(\alpha-\tilde{y})\tilde{y}\tilde{x}_2}{1-\tilde{x}_3} \end{bmatrix}, \quad \tilde{g}(\tilde{x}) = \begin{bmatrix} 0 \\ 0 \\ 0 \\ 0 \\ \dfrac{1}{1-\tilde{x}_3} \end{bmatrix}$$

On the basis of this system, one can easily verify that assumptions 1, 2 and 3 of theorem 2.1 are satisfied. A direct application of the synthesis formula of Eq.2.39 then yields the following nonlinear output feedback controllers:

$$\begin{aligned} \frac{d\omega_0}{dt} &= -\omega_0 + (1-\omega_3)Dae^{\frac{-F}{\omega_4^2}} + L_0(\tilde{h}(\tilde{x}) - \tilde{h}(\omega)) \\ \frac{d\omega_1}{dt} &= -\omega_1 + \omega_4\omega_0 + L_1(\tilde{h}(\tilde{x}) - \tilde{h}(\omega)) \\ \frac{d\omega_2}{dt} &= -\omega_2 + \omega_4\omega_1 + L_2(\tilde{h}(\tilde{x}) - \tilde{h}(\omega)) \\ \frac{d\omega_3}{dt} &= -\omega_3 + \omega_4\omega_2 + L_3(\tilde{h}(\tilde{x}) - \tilde{h}(\omega)) \\ \frac{d\omega_4}{dt} &= \frac{1-\omega_4-(\alpha-\omega_4)\omega_4\omega_2}{1-\omega_3} + L_4(\tilde{h}(\tilde{x}) - \tilde{h}(\omega)) \\ &+ \frac{[\beta_2 L_{\tilde{g}} L_{\tilde{f}}\tilde{h}(\omega)]^{-1}\left\{v - \beta_0\tilde{h}(\omega) - \beta_1 L_{\tilde{f}}\tilde{h}(\omega) - \beta_2 L_{\tilde{f}}^2\tilde{h}(\omega)\right\}}{1-\omega_3} \\ \bar{u}(t) &= [\beta_2 L_{\tilde{g}} L_{\tilde{f}}\tilde{h}(\omega)]^{-1}\left\{v - \beta_0\tilde{h}(\omega) - \beta_1 L_{\tilde{f}}\tilde{h}(\omega) - \beta_2 L_{\tilde{f}}^2\tilde{h}(\omega)\right\} \end{aligned} \tag{3.3}$$

where v is the set-point, β_0, β_1, β_2 and $L = [L_0\ L_1\ L_2\ L_3\ L_4]^T$ are controller parameters and $\tilde{h}(\omega) = \omega_0$ or $\tilde{h}(\omega) = \omega_1$. The nonlinear controller of Eq.3.3 was also combined with a PI controller (that is, the term $v - \beta_0\tilde{h}(\omega)$ was substituted by $v - \beta_0\tilde{h}(\tilde{x}) + \frac{1}{\tau_i'}\xi$, where $\dot{\xi} = v - \tilde{h}(\tilde{x})$, $\xi(0) = 0$ and τ_i' is the integral time constant) to ensure offsetless tracking in the presence of constant uncertainty in process parameters. The practical implementation of the nonlinear controllers of Eq.3.3 requires on-line measurements of the controlled outputs $\tilde{x}_0$ or $\tilde{x}_1$; in practice, such measurements can be obtained by using, for example, light scattering (Bohren and Huffman, 1983; Rawlings et al., 1993).

A schematic diagram of the closed-loop process is shown in Figure 3.1. A mix of solutions of different concentrations, with relative flow rates regulated by the controller in order to attain a desired inlet solute concentration, enter the crystallizer at a constant flow rate. Crystals, together with the solution, flow out of the crystallizer. Measurements of the moments of the crystal size distribution are obtained by a light scattering device and the control action is calculated by the controller, which is implemented in a computer. The crystallizer is assumed to be isothermal; this can be achieved by using a controller which measures the crystallizer temperature and manipulated the flow rate of the cooling water in the jacket.

1.2 Closed-loop Simulations

Several simulation runs were performed to evaluate the performance and robustness properties of the nonlinear controllers of Eq.3.3 and compare them with the ones of a PI controller. The values of the nonlinear controller parameters β_0, β_1, β_2 and L and the PI controller parameters K_c, τ_I, which were used in the simulations, are given in Tables 3.1 and 3.2 (K_c, τ_I were computed through extensive trial and error). In all the

	without disturbance and delay	with disturbance	with delay
K_c	2.5	2.5	2.5
τ_i	0.5	0.5	1.0
L	$[1.0\ 0.0\ 0.0\ 0.0\ 1.0]^T$	$[1.0\ 0.0\ 0.0\ 0.0\ 1.0]^T$	$[1.0\ 0.0\ 0.0\ 0.0\ 1.0]^T$
β_0	1.0	1.0	1.0
β_1	2/3	2/3	2/3
β_2	1/9	1/9	1/9
τ_i'	20.0	20.0	20.0

Table 3.1. Controller parameters with $\tilde{x}_0$ as controlled output.

simulation runs, the initial condition:

$$n(r, 0) = 0.0,\ c(0) = 990.0\ kg/m^3$$

was used for the process model of Eq.2.8 and the finite difference method with 1,000 discretization points was used for its simulation. The initial conditions for the dynamic system included in the controller of Eq.3.3 were set to be: $\omega_0 = 0.047$, $\omega_1 = 0.028$, $\omega_2 = 0.017$, $\omega_3 = 0.01$, and $\omega_4 = 0.5996$ (note that they do not correspond to the initial conditions used for the distributed parameter model in order to study the performance of the controller in the presence of significant initialization errors).

	without disturbance and delay	with disturbance	with delay
K_c	2.5	2.5	2.0
τ_i	0.5	0.5	0.5
L	$[1.0\ 0.0\ 0.0\ 0.0\ 1.0]^T$	$[1.0\ 0.0\ 0.0\ 0.0\ 1.0]^T$	$[1.0\ 0.0\ 0.0\ 0.0\ 1.0]^T$
β_0	1.0	1.0	1.0
β_1	2.0	2.0	2.0
β_2	1.0	1.0	1.0
$\tau_i^{'}$	20.0	20.0	20.0

Table 3.2. Controller parameters with $\tilde{x}_1$ as controlled output.

In the first set of simulation runs, $\tilde{x}_0$ was considered to be the controlled output. Initially, the set-point tracking capability of the nonlinear controller was evaluated under nominal conditions for a 0.5 increase in the value of the set-point ($v = 0.5$). Figure 3.2 shows the closed-loop output (top plot) and manipulated input (bottom plot) profiles obtained by using the nonlinear controller (solid lines) of Eq.3.3. For the sake of comparison, the corresponding profiles under PI control are also included (dashed lines). Clearly, the nonlinear controller drives the controlled output to its new set-point value in a significantly shorter time than the one required by the PI controller (note that both controlled outputs exhibit the same overshoot). For the same simulation run, the evolution of the closed-loop profile and the final steady-state profile of the crystal size distribution are shown in Figure 3.3. An exponentially-decaying crystal size distribution is obtained at the steady-state. Next, the robustness properties of the nonlinear controller in the presence of parametric uncertainties, unmodeled dynamics, and measurement sensor dead-time were investigated for a 0.5 increase in the value of the set-point. Figures 3.4 and 3.5 show the closed-loop output (top plot) and manipulated input (bottom plot) profiles under the nonlinear controller (solid lines) in the presence of 5% error in both F and τ, and in the presence of fast actuator dynamics. To account for the actuator dynamics, the process model of Eq.2.6 was augmented with the dynamical system $\epsilon_z \dot{z}_1 = -z_1 + z_2$, $\epsilon_z \dot{z}_2 = -z_2 + u$, where $z_1, z_2 \in \mathbb{R}$ are the actuator states, z_1 is the actuator output and ϵ_z is a small parameter characterizing how fast are the actuator dynamics. The corresponding output and input profiles under PI control are also included (dashed lines). In the case of parametric uncertainties, the nonlinear controller exhibits very good robustness properties, driving quickly the output to its new set-point. In the case of unmodeled actuator dynamics, the nonlinear controller was also found

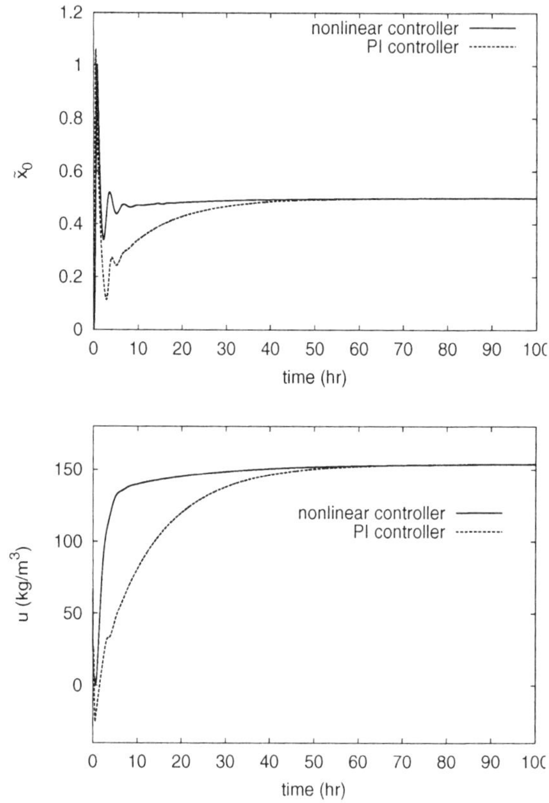

Figure 3.2. Closed-loop output (top) and manipulated input (bottom) profiles under nonlinear and PI control, for a 0.5 increase in the set-point ($\tilde{x}_0$ is the controlled output).

to be more robust since the maximum ϵ_z, for which a stable closed-loop system was obtained, under nonlinear control is $\epsilon_z = 0.04$, while under proportional-integral control $\epsilon_z = 0.02$. Finally, a 10.0-minute delay in the measurement sensor was considered and the nonlinear controller of Eq.3.3 was redesigned within a Smith-Predictor framework, according to the results in Antoniades and Christofides, 1999 (details are omitted for brevity), to account for the presence of the delay. Figure 3.6 shows the profiles of the controlled outputs (top plot) and manipulated inputs (bottom plot) under nonlinear (solid lines) and PI (dashed lines) control, for a 0.5 increase in the value of the set-point. Clearly, the presence of the measurement delay deteriorates significantly the output response

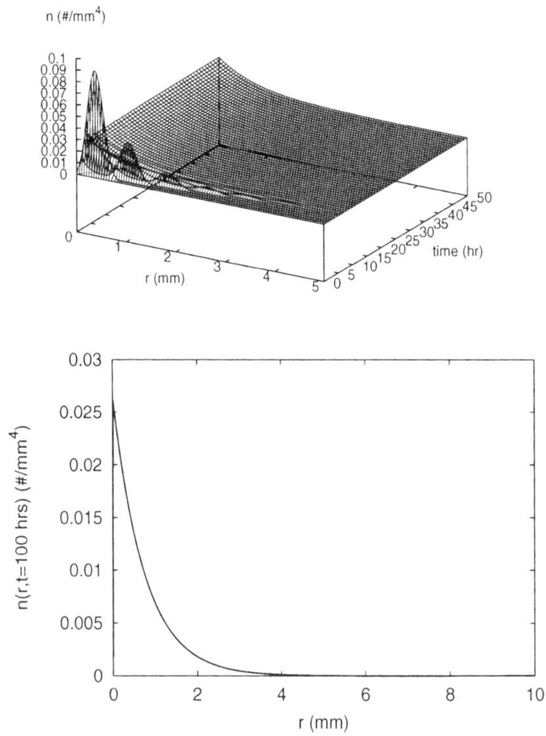

Figure 3.3. Profile of evolution of crystal size distribution (top) and final steady-state crystal size distribution (bottom) under nonlinear control ($\tilde{x}_0$ is the controlled output).

under PI control (note the oscillations of the controlled output), while it affects very little the output response under nonlinear control.

In the second set of simulation runs, $\tilde{x}_1$ was considered to be the controlled output. Initially, the performance of the nonlinear controller for a 0.5 increase in the value of the set-points was tested under nominal conditions. Figure 3.7 shows the closed-loop output (top plot) and manipulated input (bottom plot) profiles under nonlinear (solid lines) and PI (dashed lines) control. Again, the nonlinear controller drives the output to its new set-point much faster than the PI controller. The closed-loop profile of the evolution of the crystal size distribution is plotted in Figure 3.8 (top plot), along with the final steady-state profile of the crystal size distribution (middle plot) and the evolution of the mean crystal

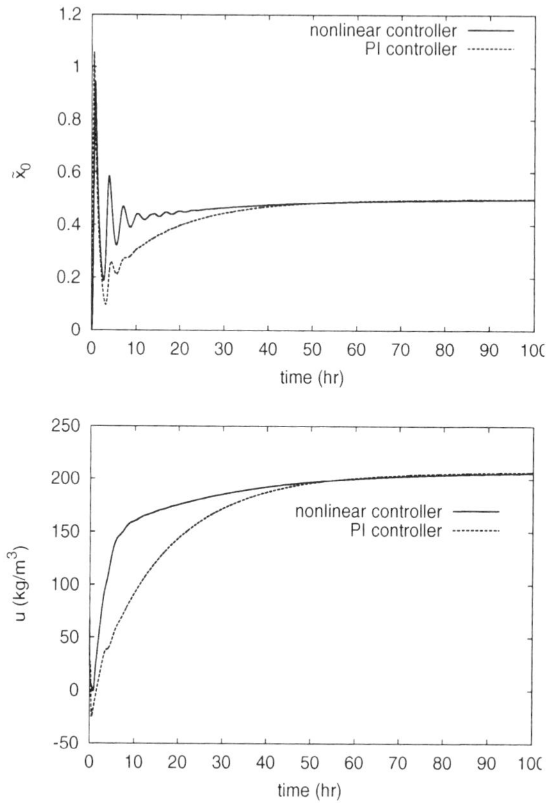

Figure 3.4. Closed-loop output (top) and manipulated input (bottom) profiles under nonlinear and PI control, for a 0.5 increase in the set-point in the presence of a 5% modeling error in both F and τ ($\tilde{x}_0$ is the controlled output).

size (bottom plot). The stabilization of the crystal size distribution is quickly achieved and an exponentially-decaying crystal size distribution is obtained at steady state. Moreover, the final steady-state mean crystal size $\tilde{x}_1/\tilde{x}_0 = 0.77$ is about 28% higher than the mean crystal size of the open-loop unstable steady-state $\tilde{x}_1/\tilde{x}_0 = 0.60$, establishing that control of $\tilde{x}_1$ allows regulating the mean crystal size at a desired value. The robustness properties of the nonlinear controller in the presence of parametric uncertainties, unmodeled dynamics, and measurement sensor dead-time were also investigated. Initially, a 5% error in both F and τ, and unmodeled actuator dynamics, as described above, were separately considered. Figures 3.9 and 3.10 show the resulting closed-loop output (top plot) and manipulated input (bottom plot) profiles under nonlinear

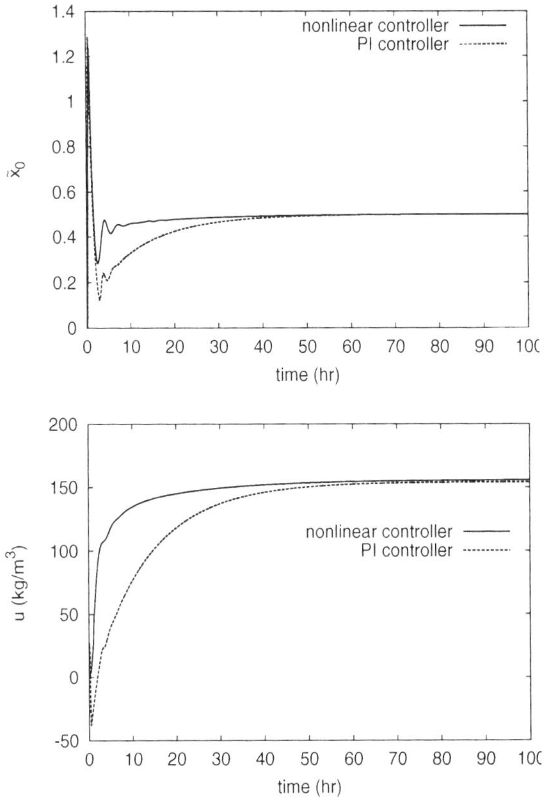

Figure 3.5. Closed-loop output (top) and manipulated input (bottom plot) profiles under nonlinear and PI control, for a 0.5 increase in the set-point in the presence of unmodeled actuator dynamics ($\tilde{x}_0$ is the controlled output).

(solid lines) and PI (dashed lines) control. It is clear that the nonlinear controller possesses very good robustness properties with respect to parametric uncertainties, attenuating their effect on the output. Also, the nonlinear controller was found to be more robust with respect to unmodeled dynamics, $\epsilon_z = 0.02$, compared to the proportional-integral controller, $\epsilon_z = 0.01$. Finally, a delay of 10.0 minutes in the output measurement was considered (again, the nonlinear controller of Eq.3.3 was redesigned within a Smith-Predictor framework to account for the measurement delay). Figure 3.11 presents the output (top plot) and manipulated input (bottom plot) profiles under nonlinear (solid lines) and PI (dashed lines) control. Again, the superior performance of the nonlinear controller is evident.

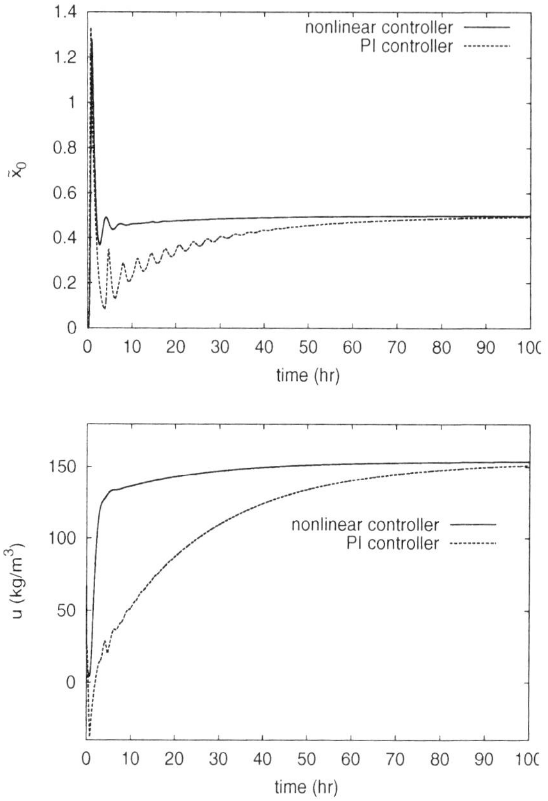

Figure 3.6. Closed-loop output (top) and manipulated input (bottom) profiles under nonlinear and PI control for a 0.5 increase in the set-point in the presence of a 10.0 minute delay in the output measurements ($\tilde{x}_0$ is the controlled output).

Remark 3.1: We note that the fifth-order model of Eq.2.14, which was used for the design of the nonlinear output feedback controllers of Eq.3.3, was obtained by using the method of moments and no improvement of its accuracy was pursued by using approximate inertial manifolds. The reason is that the closed-loop performance and robustness properties of the controllers of Eq.3.3 are clearly excellent (see the closed-loop output profiles in Figures 3.2 and 3.7, respectively), thereby leaving no room for further improvement of the performance of the controller by using approximate inertial manifolds.

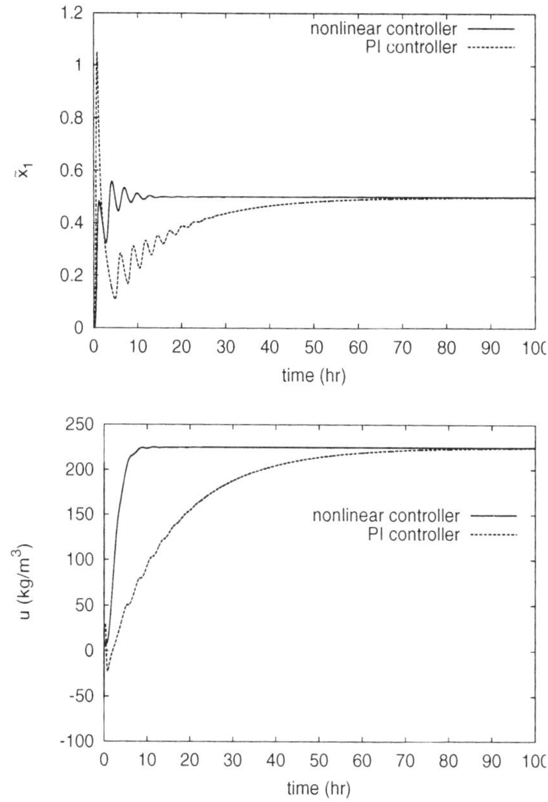

Figure 3.7. Closed-loop output (top) and manipulated input (bottom) profiles under nonlinear and PI control, for a 0.5 increase in the set-point ($\tilde{x}_1$ is the controlled output).

2. Application to a Batch Crystallizer

2.1 Introduction

A wide variety of high value products are produced through various classes of particulate processes, such as crystallization from solution. These processes are often carried out by batch cooling well-mixed crystallizers in the industry for low volume, high quality production.

Previous efforts have mainly focused on open-loop optimal control of batch cooling crystallizers (Jones, 1974; Jones and Mullin, 1974; Miller and Rawlings, 1994), and research on nonlinear feedback control of batch crystallizers has been very limited. The importance of controlling the shape of PSDs to ensure high quality production, together with recent advancements in technology for real-time PSD measurements, motivates

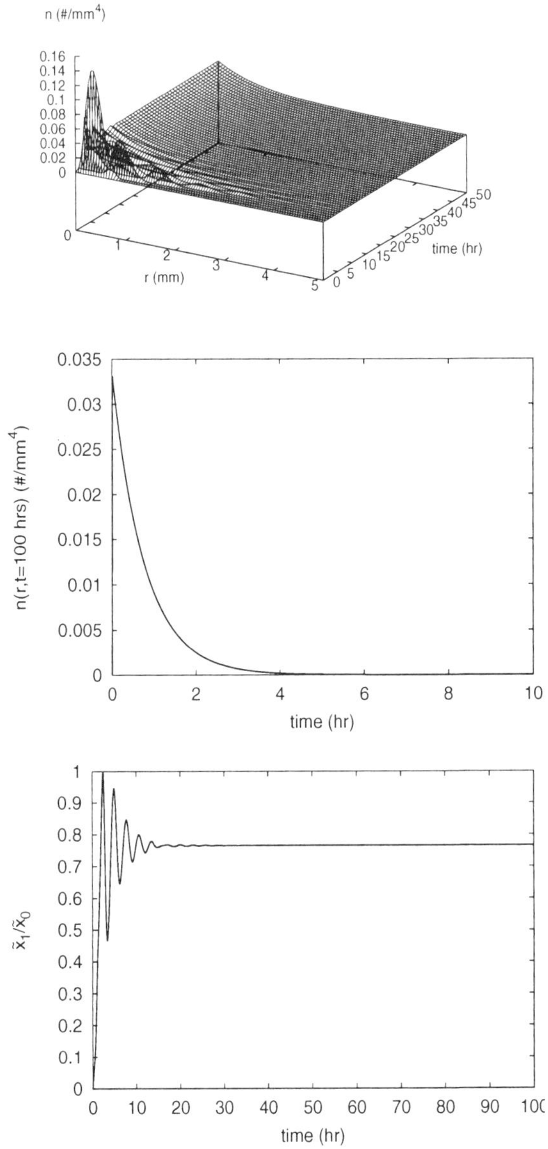

Figure 3.8. Profile of evolution of crystal size distribution (top), final steady-state crystal size distribution (middle) and evolution of mean crystal size (bottom) under nonlinear control ($\tilde{x}_1$ is the controlled output).

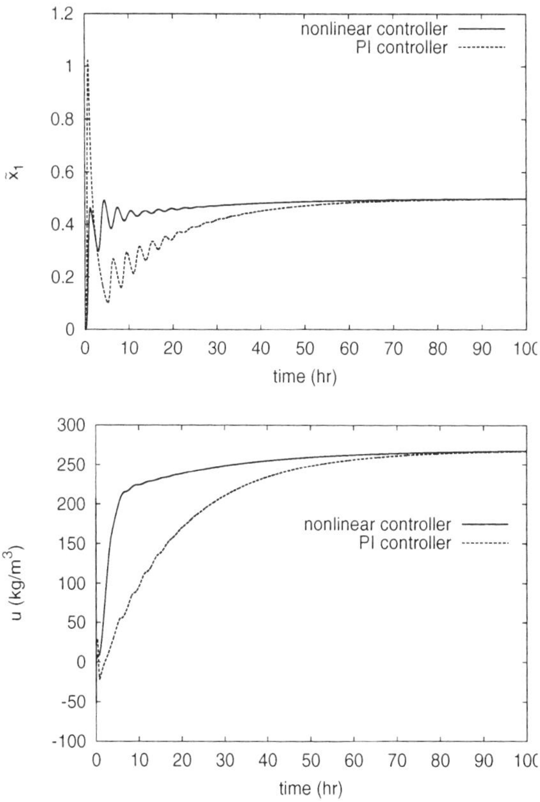

Figure 3.9. Closed-loop output (top) and manipulated input (bottom) profiles under nonlinear and PI control, for a 0.5 increase in the set-point in the presence of a 5% modeling error in both F and τ ($\tilde{x}_1$ is the controlled output).

the development of general nonlinear feedback control algorithms for batch crystallizers to enforce the desired set-point tracking properties in order to meet the necessary product specifications.

2.2 Process Modeling - Control Problem Formulation

In this section, the control method proposed in Chapter 2 is applied to a non-isothermal batch crystallizer, shown in Figure 3.12. Using similar assumptions to the one employed in the modeling of the continuous crystallizer in Chapter 2 (i.e., constant volume, mixed suspension, nucleation of crystals of infinitesimal size and mixed product removal), the

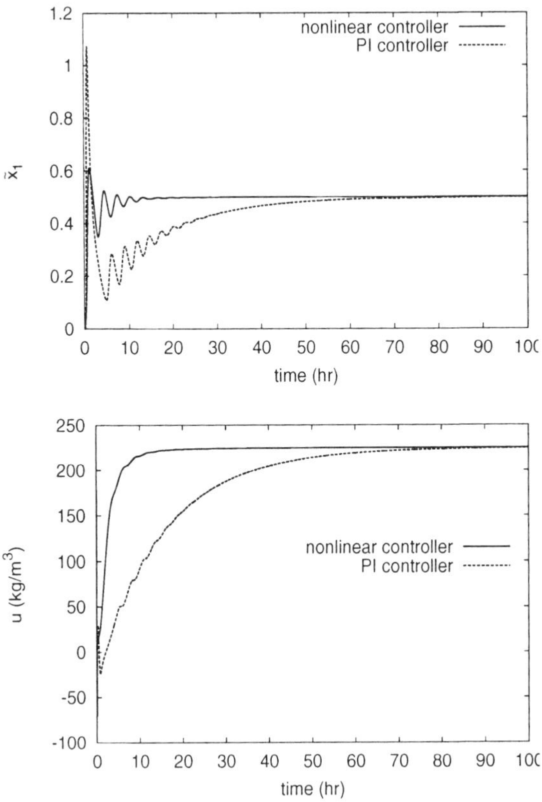

Figure 3.10. Closed-loop output (top) and manipulated input (bottom) profiles under nonlinear and PI control, for a 0.5 increase in the set-point in the presence of unmodeled actuator dynamics ($\tilde{x}_1$ is the controlled output).

model of the batch crystallization process takes the following form:

$$\begin{aligned}\frac{\partial n}{\partial t} &= -\frac{\partial(R(t)n)}{\partial r} + \delta(r-0)Q(t) \\ \frac{dc}{dt} &= \frac{(\rho - c)}{\bar{\epsilon}}\frac{d\bar{\epsilon}}{dt} \\ \frac{dT}{dt} &= \frac{\rho H_c}{\rho C_p}\frac{d\bar{\epsilon}}{dt} - \frac{U_c A_c}{\rho C_p V}(T - T_c)\end{aligned} \tag{3.4}$$

where $n(r,t)$ is the number of crystals of radius r at time t per unit volume of suspension, c is the solute concentration, c_0 is the solute concentration in the feed, $\bar{\epsilon} = 1 - \int_0^\infty n(r,t)\frac{4}{3}\pi r^3 dr$ is the volume of liquid

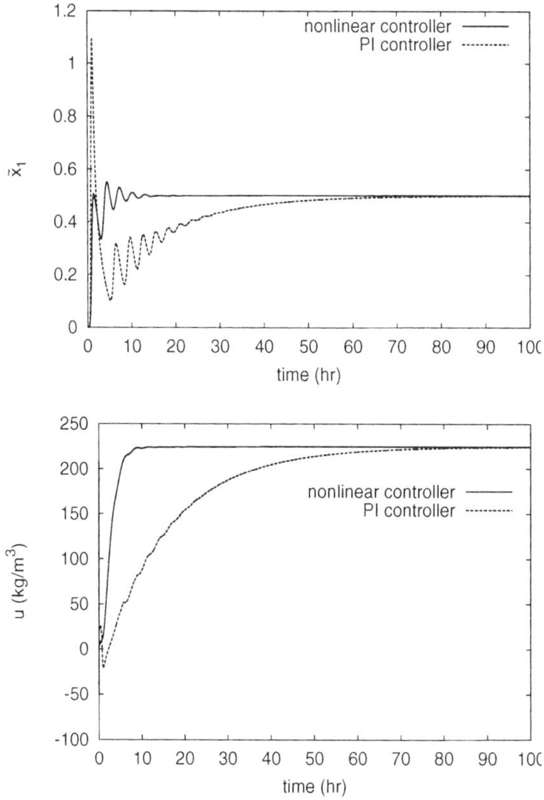

Figure 3.11. Closed-loop output (top) and manipulated input (bottom) profiles under nonlinear and PI control, for a 0.5 increase in the set-point in the presence of a 10.0-minute delay in the output measurements ($\tilde{x}_1$ is the controlled output).

per unit volume of suspension, T is the reactor temperature, H_c is the enthalpy of crystallization, C_p is the specific heat capacity, U_c is the heat transfer coefficient between the crystallizer and the cooling water, A_c is the surface area of heat transfer between the crystallizer and the cooling water, V is the reactor volume, and T_c is the cooling water temperature. $R(t)$ is the growth rate, $\delta(r-0)$ is the standard Dirac function and $Q(t)$ is the nucleation rate. The term $\delta(r-0)Q(t)$ accounts for the production of crystals of infinitesimal (zero) size via nucleation. $R(t)$ and $Q(t)$ are governed by the following relationships:

$$R(t) = k_1(c - c_s)e^{-\frac{E_g}{RT}}, \quad Q(t) = \bar{\epsilon}k_2 e^{-\frac{k_3}{\left(\frac{c}{c_s} - 1\right)^2}} e^{-\frac{E_b}{RT}} \qquad (3.5)$$

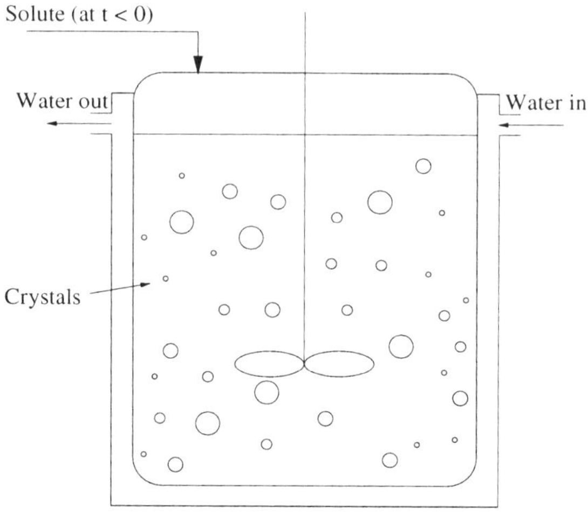

Figure 3.12. A Batch Cooling Crystallizer.

where k_1,k_2,and k_3 are constants, c_s is the concentration of solute at saturation, and E_g and E_b denote the activation energies for growth and nucleation, respectively. While k_1, k_2, k_3, E_g, E_b are taken to be constant (see Table 3.3 for their values), it is assumed that the saturation concentration, c_s, is a function of the crystallizer temperature and obeys the following relationship:

$$c_s = -3\bar{T}^2 + 38\bar{T} + 964.9 \tag{3.6}$$

where $\bar{T}$ is the dimensionless temperature defined as:

$$\bar{T} = \frac{T-273}{333-273} \tag{3.7}$$

Using the expressions of Eq.3.5, the system of Eq.3.4 can be equivalently written as:

$$\begin{aligned}
\frac{\partial n}{\partial t} &= -k_1(c-c_s)\frac{\partial n}{\partial r}e^{-\frac{E_g}{RT}} + \delta(r-0)\bar{\epsilon}k_2 e^{-\frac{k_3}{\left(\frac{c}{c_s}-1\right)^2}} e^{-\frac{E_b}{RT}} \\
\frac{dc}{dt} &= \frac{(\rho-c)}{\bar{\epsilon}}\frac{d\bar{\epsilon}}{dt} \\
\frac{dT}{dt} &= \frac{\rho H_c}{\rho C_p}\frac{d\bar{\epsilon}}{dt} - \frac{U_c A_c}{\rho C_p V}(T-T_c)
\end{aligned} \tag{3.8}$$

The control problem is formulated as the one of manipulating the deviation of the cooling water temperature ($u(t) = T_c - T_{c_s}$) to control the particle nucleation rate ($y(t) = Q$) such that a crystal size distribution with a larger average particle size is obtained at the end of the batch run. The idea behind this particular choice of control problem is to manipulate the nucleation rate such that at the end of the batch run, the average crystal size is maximized. There are many situations in industrial applications where it is desirable to obtain large crystals for easier downstream processing. It will be shown in section 2.4 that by carefully controlling the nucleation rate, especially at the early stage of the batch run, it is possible to prevent an initial "burst" of infinitesimal size crystals, which results under a linear cooling input profile and dramatically reduces the average size of crystals at the end of the batch cycle.

2.3 Nonlinear Controller Synthesis

In this section, we use the control method proposed in Chapter 2 to synthesize a nonlinear low-order controller for the batch crystallizer. The method of moments is initially used to derive a low-order ODE approximation of the population balance model of Eq.3.4, which will be used for controller design. Using the definition of moments in Eq.2.9, multiplying the population balance in Eq.3.4 by r^ν, and integrating over all particle sizes, the following system of infinite ODEs, which describes the rate of change of the moments of the particle size distribution, the solute concentration and temperature, is obtained:

$$
\begin{aligned}
\frac{d\mu_0}{dt} &= \left(1 - \frac{4}{3}\pi\mu_3\right) k_2 e^{-\frac{k_3}{\left(\frac{c}{c_s} - 1\right)^2}} e^{-\frac{E_b}{RT}} \\
\frac{d\mu_\nu}{dt} &= \nu k_1 (c - c_s) e^{-\frac{E_g}{RT}} \mu_{\nu-1}, \quad \nu = 1, 2, 3, \ldots, \\
\frac{dc}{dt} &= \frac{-4\pi(c - c_s)\mu_2(\rho - c)}{\left(1 - \frac{4}{3}\pi\mu_3\right)} \\
\frac{dT}{dt} &= -\frac{\rho_c \Delta H_c}{\rho C_p} 4\pi k_1 (c - c_s) e^{-\frac{E_g}{RT}} \mu_2 - \frac{U A_c}{\rho C_p V}(T - T_c)
\end{aligned}
\tag{3.9}
$$

On the basis of the system of Eq.3.9, it is clear that the moments of order four and higher do not affect those of order three and lower, which means that the dominant dynamics of the system of Eq.3.9 can be adequately captured by the following sixth-order model, which consists of the first four moment equations and the energy and solute mass balances:

$$
\begin{aligned}
\frac{d\mu_0}{dt} &= \left(1-\frac{4}{3}\pi\mu_3\right)k_2 e^{-\frac{k_3}{\left(\frac{c}{c_s}-1\right)^2}} e^{-\frac{E_b}{RT}}\\
\frac{d\mu_1}{dt} &= k_1(c-c_s)e^{-\frac{E_g}{RT}}\mu_0\\
\frac{d\mu_2}{dt} &= 2k_1(c-c_s)e^{-\frac{E_g}{RT}}\mu_1\\
\frac{d\mu_3}{dt} &= 3k_1(c-c_s)e^{-\frac{E_g}{RT}}\mu_2\\
\frac{dc}{dt} &= \frac{-4\pi(c-c_s)\mu_2(\rho-c)}{\left(1-\frac{4}{3}\pi\mu_3\right)}\\
\frac{dT}{dt} &= -\frac{\rho_c\Delta H_c}{\rho C_p}4\pi k_1(c-c_s)e^{-\frac{E_g}{RT}}\mu_2-\frac{UA_c}{\rho C_p V}(T-T_c)
\end{aligned}
\tag{3.10}
$$

We will now employ the nonlinear control method proposed in Chapter 2 to design a nonlinear controller for the batch crystallizer. Including the specifications for the manipulated input and the controlled output in the model of Eq.3.10, one can derive a system of the form of Eq.2.25 with $\tilde{x} = [\mu_0\ \mu_1\ \mu_2\ \mu_3\ c\ T]^T$ and:

$$
\tilde{f}(\tilde{x}) = \begin{bmatrix}
\left(1-\frac{4}{3}\pi\mu_3\right)k_2 e^{-\frac{k_3}{\left(\frac{c}{c_s}-1\right)^2}} e^{-\frac{E_b}{RT}}\\
k_1(c-c_s)e^{-\frac{E_g}{RT}}\mu_0\\
2k_1(c-c_s)e^{-\frac{E_g}{RT}}\mu_1\\
3k_1(c-c_s)e^{-\frac{E_g}{RT}}\mu_2\\
\frac{-4\pi(c-c_s)\mu_2(\rho-c)}{\left(1-\frac{4}{3}\pi\mu_3\right)}e^{-\frac{E_g}{RT}}\\
-\frac{\rho_c\Delta H_c}{\rho C_p}4\pi k_1(c-c_s)e^{-\frac{E_g}{RT}}\mu_2-\frac{UA_c}{\rho C_p V}(T-T_c)
\end{bmatrix},
$$

$$\tilde{g}(\tilde{x}) = \begin{bmatrix} 0 \\ 0 \\ 0 \\ 0 \\ 0 \\ \dfrac{UA_c}{\rho C_p V} \end{bmatrix}$$

On the basis of this system, one can easily verify that assumptions 1, 2, and 3 of theorem 2.1 are satisfied. A direct application of the synthesis formula of Eq.2.39 then yields the following nonlinear output feedback controller:

$$\begin{aligned}
\frac{d\omega_0}{dt} &= \left(1-\frac{4}{3}\pi\omega_3\right)k_2 e^{-\dfrac{k_3}{\left(\dfrac{\omega_4}{c_s}-1\right)^2}} e^{-\dfrac{E_b}{R\omega_5}} \\
\frac{d\omega_1}{dt} &= k_1(\omega_4-c_s)\omega_0 e^{-\dfrac{E_g}{R\omega_5}} + L_1(\tilde{h}(\tilde{x})-\tilde{h}(\omega)) \\
\frac{d\omega_2}{dt} &= 2k_1(\omega_4-c_s)\omega_1 e^{-\dfrac{E_g}{R\omega_5}} + L_2(\tilde{h}(\tilde{x})-\tilde{h}(\omega)) \\
\frac{d\omega_3}{dt} &= 3k_1(\omega_4-c_s)\omega_2 e^{-\dfrac{E_g}{R\omega_5}} + L_3(\tilde{h}(\tilde{x})-\tilde{h}(\omega)) \\
\frac{d\omega_4}{dt} &= \frac{-4\pi(\omega_4-c_s)\omega_2(\rho-\omega_4)}{\left(1-\dfrac{4}{3}\pi\omega_3\right)} e^{-\dfrac{E_g}{R\omega_5}} + L_4(\tilde{h}(\tilde{x})-\tilde{h}(\omega)) \\
\frac{d\omega_5}{dt} &= -\frac{\rho_c\Delta H_c}{\rho C_p}4\pi k_1(\omega_4-c_s)e^{-\dfrac{E_g}{R\omega_5}}\omega_2 - \frac{UA_c}{\rho C_p V}(\omega_5-T_c) \\
&\quad +L_5(\tilde{h}(\tilde{x})-\tilde{h}(\omega)) \\
&\quad +\frac{[\beta_2 L_{\tilde{g}} L_{\tilde{f}}\tilde{h}(\omega)]^{-1}\left\{v-\beta_0\tilde{h}(\omega)-\beta_1 L_{\tilde{f}}\tilde{h}(\omega)-\beta_2 L_{\tilde{f}}^2\tilde{h}(\omega)\right\}}{1-\omega_3} \\
\bar{u}(t) &= [\beta_2 L_{\tilde{g}} L_{\tilde{f}}\tilde{h}(\omega)]^{-1}\left\{v+\beta_1\dot{v}+\beta_2\ddot{v}-\beta_0\tilde{h}(\omega)\right. \\
&\quad \left.-\beta_1 L_{\tilde{f}}\tilde{h}(\omega)-\beta_2 L_{\tilde{f}}^2\tilde{h}(\omega)\right\}
\end{aligned} \tag{3.11}$$

where v is the set-point, β_0, β_1, β_2 and $L = [L_0\ L_1\ L_2\ L_3\ L_4]^T$ are controller parameters and $\tilde{h}(\omega) = Q(t)$.

2.4 Simulation Results

Several simulation runs were performed to evaluate the performance of the nonlinear controller of Eq.3.11 and compare with the ones of a linear input profile. The values of the controller parameters β_0, β_1, β_2, and L, which were used in the simulations, are given in Table 3.3. In all

V	=	0.01	m^3
k_1	=	5.064×10^{-2}	
k_2	=	7.957	
E_b	=	1×10^{-5}	$kJ\ mol^{-1}$
E_g	=	2200	$kJ\ mol^{-1}$
A_c	=	0.25	m^2
U	=	1800	$kJ\ m^{-2}\ hr^{-1}\ C^{-1}$
ρ_c	=	1770	$kg\ m^{-3}$
ρC_p	=	3000	$kJ\ m^{-3}\ C^{-1}$
ΔH	=	-50	$kJ\ kg^{-1}$
T_0	=	25	C
T_c	=	25	C

Table 3.3. Parameters for batch crystallizer.

simulation runs, the following initial condition was used for the process model of Eq.3.4

$$n(r,0) = 0.0, c(0) = 990\ kg/m^3 \tag{3.12}$$

and the finite difference method with 1000 discretization points was used for the simulations. The initial conditions for the observer states were computed numerically by using the initial conditions of Eq.3.12. As discussed in the continuous crystallizer example, even though the number of discretization points, 1000, used to solve the system of Eq.3.4 is very large (owing to the poor convergence properties of the finite difference scheme), the computation of an accurate (that is, independent of the discretization) solution is critical for the thorough evaluation of the performance of a nonlinear feedback controller synthesized on the basis of a low-order approximation of the distributed parameter system of Eq.3.4. The adequacy of the 1000 discretization points to yield an accurate solution was established through extensive simulations of the open- and closed-loop systems.

The reference input tracking capabilities of the nonlinear controller were evaluated with a reference trajectory of the form $v(t) = 3.5623 \times 10^{-5}e^{-1.5018t}$. Figure 3.13 shows the closed-loop output (top plot) and

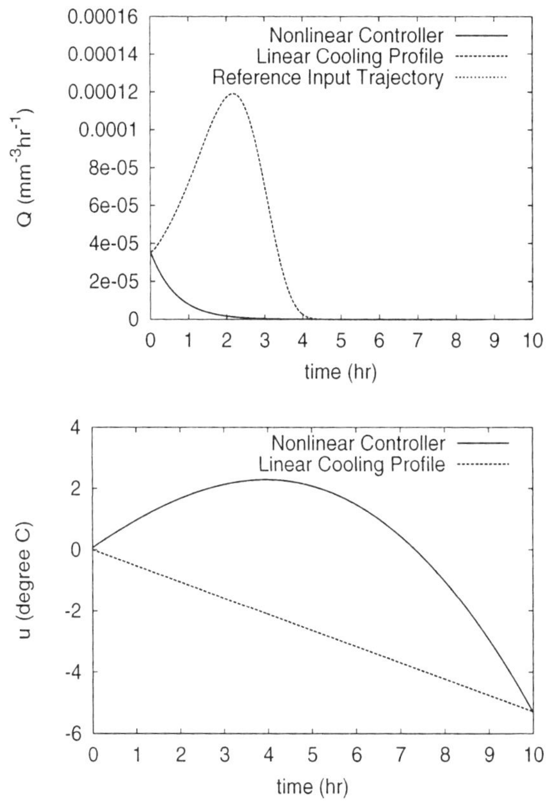

Figure 3.13. Closed-loop output (top) and manipulated input (bottom) under nonlinear control and linear cooling profile.

manipulated input (bottom plot) profiles obtained by using the nonlinear controller (solid lines) of Eq.3.11. The profiles of the nucleation rate and the reference input trajectory are shown to be almost identical. For the sake of comparison, the corresponding profiles under a linear cooling input profile of the form $u(t) = -0.528t$ are also shown in Figure 3.13. There is an initial "burst" of nucleated crystals as demonstrated by the high nucleation rate. This phenomenon affects the average particle size as can be seen in Figure 3.14. It is clear that under nonlinear control of the cooling water temperature, much larger crystals can be obtained compared to those obtained by using a linear profile for the cooling water

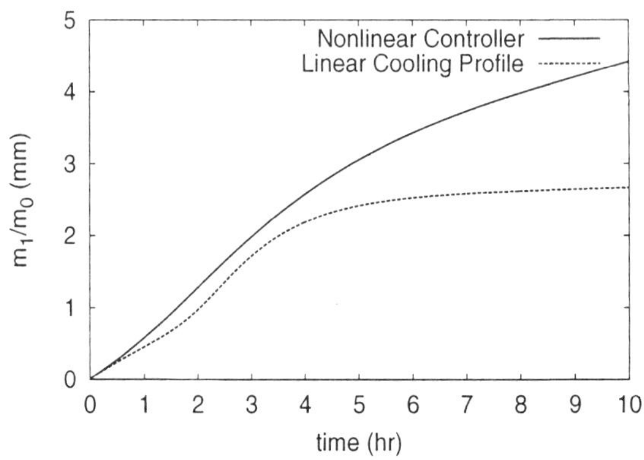

Figure 3.14. Closed-loop output under nonlinear control and linear cooling profile.

temperature. At the end of the 10-hour run, the average crystal size under the nonlinear controller of Eq.3.11 is 4.42 mm, and under the linear cooling input profile the average crystal size is 2.67 mm. This reduction in crystal size can be attributed to the fact that high nucleation rate generates a large amount of crystals of infinitesimal size, which in turn causes a significant depletion of solute, thereby prohibiting the growth of crystals through condensation of solute molecules. Figure 3.15 shows the profile of the crystal size distribution under nonlinear control (top plot) and under the linear profile for the cooling water temperature, and the final steady-state profile of the crystal size distribution are shown in Figure 3.16. It is evident that under nonlinear control, we obtain larger crystal size and smaller total number of crystals.

3. Conclusions

In this chapter, we applied the nonlinear control method proposed in Chapter 2 to continuous and batch crystallization systems. Low-dimensional approximations of the population balances were derived and used for nonlinear control design. The performance of the nonlinear controllers was successfully tested through simulations using the detailed population balance models and was shown to be superior to the one of linear control schemes.

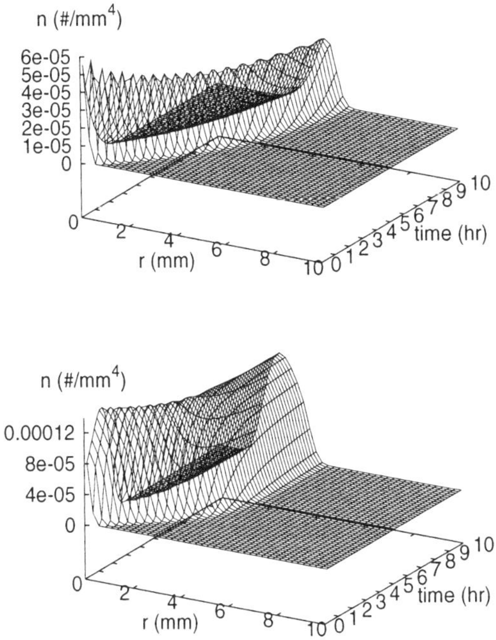

Figure 3.15. Profile of evolution of crystal size distribution under nonlinear control (top) and using a linear profile for the cooling water temperature (bottom).

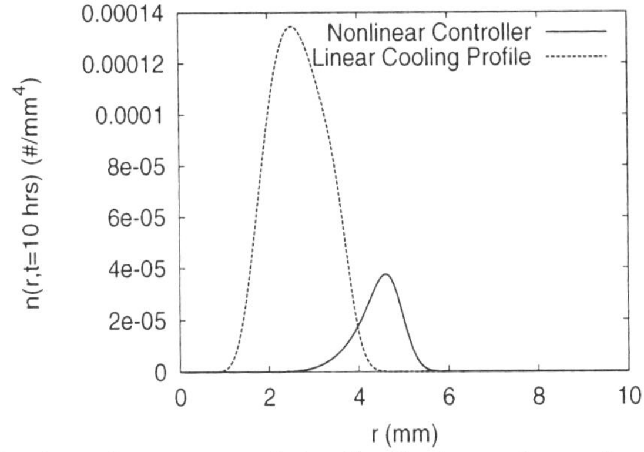

Figure 3.16. Final steady-state crystal size distribution under nonlinear control (solid line) and using a linear profile for the cooling water temperature (dashed line).

Chapter 4

ROBUST CONTROL OF PARTICULATE PROCESSES

1. Introduction

This chapter focuses on robust control of particulate processes described by uncertain population balances. The objective is to develop a general method for the synthesis of practically-implementable robust nonlinear controllers that explicitly handle time-varying uncertain variables (e.g., unknown process parameters and disturbances) and unmodeled dynamics (e.g., fast actuator and sensor dynamics not included in the process model). The robust nonlinear controllers enforce stability in the closed-loop system and attenuation of the effect of uncertain variables on the outputs, and achieve particle size distributions with desired characteristics (e.g., PSDs with desired total mass, mean particle size).

This chapter is structured as follows: After introducing the general class of uncertain particulate process models considered in this work, the method of weighted residuals is used for the construction of finite-dimensional systems that accurately reproduce the dominant dynamics of the particulate process. These ODE systems are subsequently used for the synthesis, via Lyapunov's direct method, of robust nonlinear controllers that enforce stability in the closed-loop system, attenuation of the effect of uncertain variables and achieve particle size distributions with desired characteristics. The problem of robustness with respect to unmodeled dynamics is addressed within the singular perturbation framework. It is established that the proposed robust controllers enforce the desired stability and performance specifications in the closed-loop system, provided that the unmodeled dynamics are stable and sufficiently fast. Finally, the proposed robust nonlinear control method is applied to a continuous crystallizer with fines trap, in which the nucleation rate and the crystal density change with time and the actuator and sensor dynamics are not included in the model used for the synthesis of the controller. The results of this chapter were first presented in Chiu and Christofides, 1999b; Chiu and Christofides, 2000.

2. Preliminaries

2.1 Particulate Process Model with Uncertainty

The mathematical models of particulate processes are typically obtained from an application of a population balance to the particle phase which accounts for particle growth, nucleation, agglomeration and breakage, as well as from the application of material and energy balances to the continuous phase. Uncertainty in particulate process models arises from two sources: uncertain variables (e.g., unknown process parameters and external disturbances) and unmodeled dynamics (e.g., fast actuator and sensor dynamics which are not taken into account in the process model). In order to develop a general control method for particulate processes with uncertain variables and unmodeled dynamics, we consider the following singularly perturbed system of nonlinear partial integro-differential equations:

$$
\begin{aligned}
\frac{\partial n}{\partial t} &= -\frac{\partial(G(x,r)n)}{\partial r} + \bar{w}(n,x,r,z,\theta(t)) + \bar{g}_1(n,x,r)u(t), \\
& \quad n(0,t) = b(x(t)) \\
\dot{x} &= \bar{f}(x) + Q_1(x)z + \bar{g}_2(x)u(t) + \bar{g}_3(x,\theta(t), \int_0^{r_{max}} a_2(n,r,x)dr) \\
\epsilon\dot{z} &= \tilde{f}(x) + Q_2(x)z + \tilde{g}_2(x)u(t) + \tilde{g}_3(x,\theta(t), \int_0^{r_{max}} a_2(n,r,x)dr)
\end{aligned}
\tag{4.1}
$$

where $n(r,t) \in L_2[0,r_{max})$ is the size distribution function, ($L_2[0,r_{max})$ is the Hilbert space of continuous functions defined on the interval $[0,r_{max})$), $r \in [0,r_{max}]$ is the particle size and r_{max} is the maximum particle size (which may be infinity), t is the time and $x \in \mathbb{R}^n$ is the vector of state variables which describe properties of the continuous phase (e.g., solute concentration, temperature and pH in a crystallizer), $u(t) = [u_1 \ u_2 \ \cdots \ u_m]^T \in \mathbb{R}^m$ is the vector of manipulated inputs, $\theta(t) = [\theta_1 \ \theta_2 \ \cdots \ \theta_q]^T \in \mathbb{R}^q$ denotes the vector of uncertain variables, $z \in \mathbb{R}^p$ is the vector of the fast (unmodeled) process dynamics, and ϵ is a small positive parameter which quantifies the speed ratio of the slow versus the fast dynamical phenomena of the process. $(G,\bar{w})$, $(\bar{g}_1,\bar{f},\bar{g}_3,\tilde{f},\tilde{g}_3,a_2)$, and $(Q_1,\bar{g}_2,Q_2,\tilde{g}_2)$ are nonlinear smooth scalar functions, vectors and matrices, respectively.

In Eq.4.1, the n-equation is the population balance where $G(x, r)$ is the growth rate and accounts for particle growth through condensation, and $w(n, x, r, z, \theta(t))$ is a term that accounts for the net rate of introduction of new particles into the system (it includes all the means by which particles appear or disappear within the system including particle agglomeration, breakage, nucleation, feed and removal). The x-subsystem of Eq.4.1 is derived by applying material and energy balances to the continuous phase, while the z-subsystem of Eq.4.1 represents the fast dynamics which are present in the process but are neglected in the model used for controller design. Finally, the terms $\bar{g}_3(x, \theta(t), \int_0^{r_{max}} a_2(n, r, x)dr)$ and $\tilde{g}_3(x, \theta(t), \int_0^{r_{max}} a_2(n, r, x)dr)$ account for mass and heat transfer from the continuous phase to all the particles in the population.

We define a vector of controlled outputs to express the various control objectives (e.g. regulation of total number of particles, mean particle size, temperature, pH, etc.) as:

$$y_i(t) = h_i(\int_0^{r_{max}} c_\kappa(r)n(r, t)dr, x), \quad i = 1, \ldots, m, \quad \kappa = 1, \ldots, l \tag{4.2}$$

where $y_i(t)$ is the ith controlled output, $h_i(\int_0^{r_{max}} c_\kappa(r)n(r, t)dr, x)$ is a nonlinear scalar smooth function of its arguments and $c_\kappa(r)$ is a known smooth function of r which depends on the desired performance specifications.

Remark 4.1: Referring to the general mathematical model of Eq.4.1, we note that the vector of uncertain variables, $\theta(t)$, and the vector of manipulated inputs, $u(t)$, appear in all the equations of the model; this will allow addressing control problems in which manipulated inputs appear in the population balance equation (see example of crystallizer with fines trap for a demonstration of this point below).

Remark 4.2: The derivation of a singularly perturbed representation of a nonlinear process that exhibits two-time-scale behavior is, in general, a highly nontrivial task. The natural approach to address this problem involves defining the singular perturbation parameter ϵ, taking into account the physicochemical characteristics of the process, so that in the resulting singularly perturbed representation the separation of the fast and slow variables is consistent with the process dynamic behavior. This approach works for the majority of two-time-scale processes (see for example the applications considered in Kokotovic et al., 1986). Whenever this approach does not work alternative approaches that utilize explicit coordinate changes (e.g., Kokotovic et al., 1986; Kumar et al., 1998) may be employed to derive a singularly perturbed representation of a two-time-scale process. Referring to the specific singularly perturbed system of Eq.4.1, we note that the parameter ϵ appears only in the left-hand side (multiplying the time-derivative $\dot{z}$), while the fast variable z enters in an affine fashion. The first assumption is made for notational simplicity and can be readily relaxed, while the second assumption is consistent with the fact that in many physical and chemical processes the main nonlinearities are associated with the slow dynamics.

2.2 Two-time-scale Analysis

The central idea of the two-time-scale analysis is to infer the stability properties of and synthesize well-conditioned nonlinear controllers for the singularly perturbed system of Eq.4.1 based on ϵ-independent models that describe the slow and fast dynamics of this system in the slow and fast time-scale, respectively. Setting $\epsilon = 0$ in the system of Eq.4.1 and assuming that $Q_2(x, \theta)$ is invertible uniformly in $x \in \mathbb{R}^n$, $\theta \in \mathbb{R}^q$, the following system which describes the slow dynamics of the system of Eq.4.1 (called slow subsystem) is obtained:

$$
\begin{aligned}
\frac{\partial n}{\partial t} &= -\frac{\partial(G(x,r)n)}{\partial r} + w(n,x,r,\theta(t)) + g_1(n,x,r)u(t), \\
& \quad n(0,t) = b(x(t)) \\
\dot{x} &= f(x) + g_2(x)u(t) + g_3(x, \theta(t), \int_0^{r_{max}} a_2(n,r,x)dr)
\end{aligned}
\tag{4.3}
$$

where f, a_2, g_1, g_3, g_2 are nonlinear functions whose explicit form is omitted for brevity. The system which describes the fast dynamics of the system of Eq.4.1 (called fast subsystem) can be obtained by defining the fast time-scale $\tau = \frac{t}{\epsilon}$, deriving the representation of the system of Eq.4.1 in the τ time scale and setting $\epsilon = 0$, and is of the form:

$$
\frac{dz}{d\tau} = \tilde{f}(x) + Q_2(x)z + \tilde{g}_2(x)u(t) + \tilde{g}_3(x, \theta(t), \int_0^{r_{max}} a_2(n,r,x)dr) \tag{4.4}
$$

where x and n can be considered equal to their initial values $x(0)$ and $n(r, 0)$, and θ can be viewed as constant. To simplify the development of the theoretical results of the manuscript, we consider systems of the form of Eq.4.1, for which the corresponding fast subsystem of the form of Eq.4.4 is globally asymptotically stable (i.e., the eigenvalues of the matrix $Q_2(x)$ lie in the left half of the complex plane uniformly in $x \in \mathbb{R}^n$).

Remark 4.3: Whenever the open-loop fast subsystem of Eq.4.1 is unstable (i.e., one of the eigenvalues of the matrix $Q_2(x)$ lies in the right-half of the complex plane) and the pair $[Q_2 \; \tilde{g}_2]$ is stabilizable, a preliminary state feedback law of the form:

$$u = k^T(x)z + \tilde{u} \tag{4.5}$$

where $\tilde{u}$ is an auxiliary input, can be used to stabilize the fast dynamics, thereby yielding a two-time-scale system with stable fast dynamics. The design of the gain $k^T(x)$ can be performed by using standard optimal control methods (see, for example, Kokotovic et al., 1986).

3. Robust Nonlinear Control of Particulate Processes

The objective of this section is to synthesize robust nonlinear controllers for particulate processes of the form of Eq.4.3 that enforce stability and robust output tracking in the closed-loop system. Owing to the fact that the unmodeled dynamics in the model of Eq.4.1 are stable, the controllers will be synthesized on the basis of the model of Eq.4.3. Since this model is infinite-dimensional, we will initially use the method of weighted residuals to derive an ODE approximation of the system of Eq.4.3 which will be used for the synthesis of the robust nonlinear controllers.

3.1 Model Reduction

We initially use the method of weighted residuals to derive a nonlinear set of ODEs that accurately reproduces the solutions and the dominant dynamics of the distributed parameter system of Eq.4.3. The central idea of the method of weighted residuals is to approximate the exact solution of $n(r,t)$ by an infinite series of orthogonal basis functions defined in the interval $[0, r_{max})$ with time-varying coefficients, substitute the series expansion into Eq.4.3, and then take the inner product with respect to a complete set of weighted functions, to compute a set of ODEs which describes the rate of change of the time-varying coefficients of the series expansion of the solution. Specifically, we expand the solution of $n(r,t)$ in an infinite series in terms of an orthogonal and complete set of basis functions, $\phi_k(r)$, where $r \in [0, r_{max})$, $k = 1, \ldots, \infty$, as follows:

$$n(r,t) \;=\; \sum_{k=1}^{\infty} a_k(t)\phi_k(r) \tag{4.6}$$

where $a_k(t)$ are time-varying coefficients. Substituting the above expansion into the particulate process model of Eq.4.3, we obtain:

$$\begin{aligned}\sum_{k=1}^{\infty}\phi_k(r)\frac{\partial a_k(t)}{\partial t} &= -\sum_{k=1}^{\infty}a_k(t)\frac{\partial(G(x,r)\phi_k(r))}{\partial r}\\ &\quad +w(\sum_{k=1}^{\infty}a_k(t)\phi_k(r),x,r,\theta(t))\\ &\quad +g_1(\sum_{k=1}^{\infty}a_k(t)\phi_k(r),x,r)u(t)\\ \dot{x} &= f(x)+g_2(x)u(t)+g_3(x,\theta(t),\int_0^{r_{max}}a_2(\sum_{k=1}^{\infty}a_k(t)\phi_k(r),r,x)dr)\end{aligned} \tag{4.7}$$

Multiplying the population balance with the weighting functions, $\psi_\nu(r)$, and integrating over the entire particle size spectrum (i.e., taking inner product in $L_2[0,r_{max})$ with the weighting functions), the following set of infinite ODEs is obtained:

$$\begin{aligned}\int_0^{r_{max}}\psi_\nu(r)\sum_{k=1}^{\infty}\phi_k(r)\frac{\partial a_k(t)}{\partial t}dr &= -\sum_{k=1}^{\infty}a_k(t)\int_0^{r_{max}}\psi_\nu(r)\frac{\partial(G(x,r)\phi_k(r))}{\partial r}dr\\ &\quad +\int_0^{r_{max}}\psi_\nu(r)w(\sum_{k=1}^{\infty}a_k(t)\phi_k(r),x,r,\theta(t))dr\\ &\quad +\int_0^{r_{max}}\psi_\nu(r)g_1(\sum_{k=1}^{\infty}a_k(t)\phi_k(r),x,r)u(t)dr,\ \ \nu=1,\ldots,\infty\\ \dot{x} &= f(x)+g_2(x)u(t)+g_3(x,\theta(t),\int_0^{r_{max}}a_2(\sum_{k=1}^{\infty}a_k(t)\phi_k(r),r,x)dr)\end{aligned} \tag{4.8}$$

Eq.4.8 is an infinite set of ODEs which describe the rate of change of the time varying coefficients, $a_k(t)$, where $k=1,\ldots,\infty$, of the series expansion of the solution. An accurate approximation of Eq.4.8 is obtained by truncating the series expansion of $n(r,t)$ up to order N and keeping the first N equations (i.e., $\nu=1,\ldots,N$). The infinite-dimensional system of Eq.4.8 reduces to the following finite set of ODEs:

$$
\begin{aligned}
&\int_0^{r_{max}} \psi_\nu(r)\sum_{k=1}^{N}\phi_k(r)\frac{\partial a_{kN}(t)}{\partial t}dr \\
&\quad = -\sum_{k=1}^{N} a_{kN}(t)\int_0^{r_{max}}\psi_\nu(r)\frac{\partial(G(x_N,r)\phi_k(r))}{\partial r}dr \\
&\quad +\int_0^{r_{max}}\psi_\nu(r)w(\sum_{k=1}^{N}a_{kN}(t)\phi_k(r),x_N,r,\theta(t))dr \\
&\quad +\int_0^{r_{max}}\psi_\nu(r)g_1(\sum_{k=1}^{N}a_{kN}(t)\phi_k(r),x_N,r)u(t)dr, \quad \nu=1,\ldots,N \\
&\dot{x}_N = f(x_N)+g_2(x_N)u(t) \\
&\quad +g_3(x_N,\theta(t),\int_0^{r_{max}}a_2(\sum_{k=1}^{N}a_{kN}(t)\phi_k(r),r,x_N)dr)
\end{aligned} \tag{4.9}
$$

where x_N and a_{kN} are the approximations of x and a_k obtained by an N-th order truncation. Introducing the vector notation $a_N = [a_{1N} \; \cdots \; a_{NN}]$, and after some rearrangements, Eq.4.9 can be represented in the following general form:

$$
\begin{aligned}
\dot{a}_N &= \bar{f}(a_N,x_N,\theta(t))+\bar{g}(a_N,x_N)u(t) \\
\dot{x}_N &= f(x_N)+g_2(x_N)u(t)+g_3(x_N,a_N,\theta(t))
\end{aligned} \tag{4.10}
$$

where the explicit expressions of $\bar{f}(a_N,x_N,\theta(t))$ and $\bar{g}(a_N,x_N)$ are omitted for brevity. Setting $\tilde{x} = [a_N^T \; x_N^T]^T$, we obtain the following system:

$$
\begin{aligned}
\dot{\tilde{x}} &= \tilde{f}(\tilde{x})+\sum_{i=1}^{m}\tilde{g}_i(\tilde{x})u_i+\tilde{w}(\tilde{x},\theta) \\
y_{si} &= \tilde{h}_i(\tilde{x}), \quad i=1,\ldots,m
\end{aligned} \tag{4.11}
$$

where $\tilde{f}(\tilde{x}), \tilde{g}_i(\tilde{x}), \tilde{w}(\tilde{x},\theta)$ are nonlinear vector functions whose explicit form is omitted brevity.

3.2 Robust Nonlinear Controller Synthesis

The objective of this section is to use the ODE system of Eq.4.11 to synthesize robust state feedback controllers of the form:

$$
u = p(\tilde{x})+Q(\tilde{x})\bar{v}+r(\tilde{x},t) \tag{4.12}
$$

where $p(\tilde{x})$, $r(\tilde{x},t)$ are vector functions, $Q(\tilde{x})$ is a matrix, and $\bar{v}$ is a vector of the form $\bar{v} = \mathcal{V}(v_i, v_i^{(1)}, \ldots, v_i^{(r_i)})$ where $\mathcal{V}(v_i, v_i^{(1)}, \ldots, v_i^{(r_i)})$ is a smooth vector function, $v_i^{(k)}$ is the kth time derivative of the external reference input v_i (which is assumed to be a smooth function of time) and r_i is a positive integer, that enforce boundedness of the states and output tracking with arbitrary degree of asymptotic attenuation of the effect of the uncertainty on the output, in the closed-loop system, provided that ϵ is sufficiently small (i.e., the unmodeled dynamics are sufficiently fast). The control law of Eq.4.12 comprises of the component $p(\tilde{x}) + Q(\tilde{x})\bar{v}$, which is responsible for the output tracking and stabilization of the closed-loop slow system, and the component $r(\tilde{x},t)$ which is responsible for the asymptotic attenuation of the effect of the uncertain variables on the outputs of the system of Eq.4.11. The control law of Eq.4.12 will be synthesized constructively using Lyapunov's direct method, and assuming the existence of known bounding functions that capture the magnitude of the uncertain terms and that certain structural conditions on the way the uncertain variables affect the output are satisfied. Finally, the reader may refer to remark 4.6 below for results on the implementation of a control law of the form of Eq.4.12 with a state observer, when measurements of the states of the system of Eq.4.1 are not available.

In order to develop a solution to the above robust control problem, we will need to impose the following three assumptions on the system of Eq.4.11. We initially assume that there exists a coordinate transformation that renders the system of Eq.4.11 partially linear. This assumption is motivated by the requirement of robust output tracking and is precisely formulated below:

Assumption 4.1 : *Referring to the system of Eq.4.11, there exist a set of integers* $(r_1, r_2, \ldots, r_m)$ *and a coordinate transformation* $(\zeta, \eta) = T(x_s, \theta)$ *such that the representation of the system, in the coordinates* (ζ, η), *takes the form:*

$$
\begin{aligned}
\dot{\zeta}_1^{(1)} &= \zeta_2^{(1)} \\
&\vdots \\
\dot{\zeta}_{r_1-1}^{(1)} &= \zeta_{r_1}^{(1)} \\
\dot{\zeta}_{r_1}^{(1)} &= L_{\tilde{f}}^{r_1}\tilde{h}_1(\tilde{x}) + \sum_{i=1}^{m} L_{\tilde{g}_i} L_{\tilde{f}}^{r_1-1}\tilde{h}_1(\tilde{x})u_i \ + L_{\tilde{w}} L_{\tilde{f}}^{r_1-1}\tilde{h}_1(\tilde{x}) \\
&\vdots
\end{aligned}
$$

$$
\begin{aligned}
\dot{\zeta}_1^{(m)} &= \zeta_2^{(m)} \\
&\vdots \\
\dot{\zeta}_{r_m-1}^{(m)} &= \zeta_{r_m}^{(l)} \\
\dot{\zeta}_{r_m}^{(m)} &= L_{\tilde{f}}^{r_m}\tilde{h}_m(\tilde{x}) + \sum_{i=1}^{m} L_{\tilde{g}_i} L_{\tilde{f}}^{r_m-1}\tilde{h}_m(\tilde{x})u_i + L_{\tilde{w}} L_{\tilde{f}}^{r_m-1}\tilde{h}_m(\tilde{x}) \\
\dot{\eta}_1 &= \Psi_1(\zeta, \eta, \theta, \dot{\theta}) \\
&\vdots \\
\dot{\eta}_{(n+N)-\sum_i r_i} &= \Psi_{(n+N)-\sum_i r_i}(\zeta, \eta, \theta, \dot{\theta})
\end{aligned}
$$

$$
y_{si} = \zeta_1^{(i)}, \quad i = 1, \dots, m \tag{4.13}
$$

where $x_s = T^{-1}(\zeta, \eta, \theta)$, $\zeta = [\zeta^{(1)} \ \cdots \ \zeta^{(m)}]^T \in \mathbb{R}^{\sum_i r_i}$, $\eta = [\eta_1 \ \cdots \ \eta_{(n+N)-\sum_i r_i}]^T \in \mathbb{R}^{(n+N)-\sum_i r_i}$

Assumption 4.1 includes the matching condition of our robust control methodology. In particular, we consider systems of the form Eq.4.11 for which the time-derivatives of the output y_{si} up to order $r_i - 1$ are independent of the vector of uncertain variables θ. Notice that this condition is different than the standard one which restricts the uncertainty vector θ to enter the system of Eq.4.11 in the same equation with the manipulated input u. The motivation for considering this matching condition is given by the fact that it is satisfied by a large number of practical applications (note that assumption 4.1 is always satisfied for systems for which $r_i = 1$, for all $i = 1, \dots, m$.)

Referring to the system of Eq.4.11, we will assume, in order to simplify the presentation of our results, that the matrix:

$$
C(\tilde{x}) = \begin{bmatrix} L_{\tilde{g}_1} L_{\tilde{f}}^{r_1-1}\tilde{h}_1(\tilde{x}) & \cdots & L_{\tilde{g}_m} L_{\tilde{f}}^{r_1-1}\tilde{h}_1(\tilde{x}) \\ \vdots & \cdots & \vdots \\ L_{\tilde{g}_1} L_{\tilde{f}}^{r_m-1}\tilde{h}_m(\tilde{x}) & \cdots & L_{\tilde{g}_m} L_{\tilde{f}}^{r_m-1}\tilde{h}_m(\tilde{x}) \end{bmatrix} \tag{4.14}
$$

is nonsingular uniformly in $\tilde{x}$. This assumption can be readily relaxed if robust dynamic state feedback, instead of robust state feedback, is used to solve the control problem (see Isidori, 1995 for details).

The next assumption is made to ensure bounded stability of the internal dynamics of the system of Eq.4.11 under a robust state feedback controller of the form of Eq.4.11.

Assumption 4.2: *The dynamical system:*

$$\begin{array}{rcl} \dot{\eta}_1 & = & \Psi_1(\zeta, 0, 0, 0) \\ & \vdots & \\ \dot{\eta}_{(n+N)-\sum_i r_i} & = & \Psi_{(n+N)-\sum_i r_i}(\zeta, 0, 0, 0) \end{array} \tag{4.15}$$

is locally exponentially stable.

Finally, we need to assume that there exist a nonlinear time-varying function that captures the size of the uncertain term $[L_{\tilde{w}} L_{\tilde{f}}^{r_1-1} \tilde{h}_1(\tilde{x}) \ \cdots \ L_{\tilde{w}} L_{\tilde{f}}^{r_m-1} \tilde{h}_m(\tilde{x})]$. Information of this kind may result from physical considerations, preliminary simulations, experimental data, etc.

Assumption 4.3: *There exists a known function $c(\tilde{x}, t)$ such that the following condition holds:*

$$|[L_{\tilde{w}} L_{\tilde{f}}^{r_1-1} \tilde{h}_1(\tilde{x}) \ \cdots \ L_{\tilde{w}} L_{\tilde{f}}^{r_m-1} \tilde{h}_m(\tilde{x})]^T| \leq c(\tilde{x}, t) \tag{4.16}$$

for all $\tilde{x} \in \mathbb{R}^{n+N}$, $\theta \in \mathbb{R}^q$, $t \geq 0$.

Theorem 4.1 that follows provides an explicit formula for the robust controller, conditions that ensure boundedness of the state, and a precise characterization of the ultimate uncertainty attenuation level. To simplify the statement of the theorem, we set $\bar{v}_i = [v_i \ v_i^{(1)} \ \cdots \ v_i^{(r_i)}]^T$ and $\bar{v} = [\bar{v}_1^T \ \bar{v}_2^T \ \cdots \ \bar{v}_m^T]^T$. The proof this result is given in Appendix 9.B.

Theorem 4.1: *Suppose that assumptions 4.1, 4.2, and 4.3 hold, and consider the system of Eq.4.1 under the robust state feedback controller:*

$$\begin{array}{rcl} u = a(\tilde{x}, \bar{v}, t) & := & [C(\tilde{x})]^{-1} \left\{ \displaystyle\sum_{i=1}^{m} \sum_{k=1}^{r_i} \frac{\beta_{ik}}{\beta_{ir_i}} (v_i^{(k)} - L_{\tilde{f}}^k \tilde{h}_i(\tilde{x})) \right. \\ & & \displaystyle + \sum_{i=1}^{m} \sum_{k=1}^{r_i} \frac{\beta_{ik}}{\beta_{ir_i}} (v_i^{(k-1)} - L_{\tilde{f}}^{k-1} \tilde{h}_i(\tilde{x})) \\ & & \left. \displaystyle - \chi[c(\tilde{x}, t)] \frac{\displaystyle\sum_{i=1}^{l} \sum_{k=1}^{r_i} \frac{\beta_{ik}}{\beta_{ir_i}} (L_{\tilde{f}}^{k-1} \tilde{h}_i(\tilde{x}) - v_i^{(k-1)})}{|\displaystyle\sum_{i=1}^{l} \sum_{k=1}^{r_i} \frac{\beta_{ik}}{\beta_{ir_i}} (L_{\tilde{f}}^{k-1} \tilde{h}_i(\tilde{x}) - v_i^{(k-1)})| + \phi} \right\} \end{array} \tag{4.17}$$

where $\dfrac{\beta_{ik}}{\beta_{ir_i}} = [\dfrac{\beta_{ik}^1}{\beta_{ir_i}^1} \ \cdots \ \dfrac{\beta_{ik}^l}{\beta_{ir_i}^l}]^T$ are column vectors of parameters chosen so that the roots of the equation $det(B(s)) = 0$, where $B(s)$ is an $m \times m$

matrix, whose (i, j)*-th element is of the form* $\sum_{k=1}^{r_i} \frac{\beta_{jk}^i}{\beta_{jr_i}^i} s^{k-1}$*, lie in the open left-half of the complex plane, and* χ, ϕ *are adjustable parameters with* $\chi > 1$ *and* $\phi > 0$*. Then, there exist positive real numbers* (δ, ϕ^*, d) *such that for each* $\phi \leq \phi^*$*, there exists* $\epsilon^*(\phi)$*, such that if* $\phi \leq \phi^*$*,* $\epsilon \leq \epsilon^*(\phi)$ *and* $\max\{|x(0)|, |z(0)|, ||\eta(r, 0)||_2, ||\theta||, ||\dot{\theta}||, ||\tilde{v}||\} \leq \delta$*,*
a) the state of the infinite-dimensional closed-loop system is bounded, and
b) the outputs of the infinite-dimensional closed-loop system satisfy:

$$\limsup_{t \to \infty} |y_i - v_i| \leq d, \quad i = 1, \ldots, l \tag{4.18}$$

Remark 4.4: Regarding the practical application of Theorem 4.1, one has to initially use the method of weighted residuals to derive an ODE system of the form of Eq.4.11, and then verify assumptions 4.1, 4.2 and 4.3 on the basis of this system. Then, the synthesis formula of Eq.4.17 can be directly used to derive the explicit form of the controller. Moreover, the value of ϵ in the model of Eq.4.1 is typically fixed by the process, say ϵ_p, and thus there is a limit on how small the ultimate bound d can be chosen. For example, one can initially compute, through simulations, a ϕ^* from the desired (δ, d) and, in turn, the value ϵ^* for $\phi \leq \phi^*$. If this ϵ^* is less than ϵ_p, then d may need to be readjusted (increased) so that $\epsilon^* \geq \epsilon_p$. Of course, if ϵ_p is too large, there may be no value of d that works.

Remark 4.5: Referring to the robust nonlinear controller of Eq.4.17, we note that the nonlinear term $-\chi[c(\tilde{x}, t)]$ could have been replaced by a sufficiently large positive constant k. Although this modification would lead to a simplification in the practical implementation of the controller, we select to use the nonlinear term, because the use of a large positive constant results in a controller that computes very large control action, when the discrepancy between y and v (tracking error) is far from being zero. The controller that uses the nonlinear term avoids this problem and does not compute unnecessarily large control action (see the manipulated input profiles for the crystallizer example presented in section 4.4).

Remark 4.6: The on-line implementation of the controller of Eq.4.17 requires that the values of the state variables $\tilde{x}$ are known. Unfortunately, $\tilde{x}$ may not be known in many practical applications. One way to address this problem is to use a nonlinear state observer of the form:

$$\frac{d\omega}{dt} = \tilde{f}(\omega) + \tilde{g}(\omega)u + \tilde{w}(\omega)\theta_n + L(y - \tilde{h}(\omega)) \tag{4.19}$$

where ω denotes the observer state vector (the dimension of the vector ω is equal to the dimension of $\tilde{x}$ in the system of Eq.4.11), $y = [y_1\ y_2\ \cdots y_m]^T$ is the measured output vector, θ_n denotes a nominal value for $\theta(t)$ and L is a matrix chosen so that the eigenvalues of the matrix $C_L = \frac{\partial \tilde{f}}{\partial \omega}_{(\omega=\omega_s)} - L\frac{\partial \tilde{h}}{\partial \omega}_{(\omega=\omega_s)}$, where ω_s is the operating steady-state, lie in the open left-half of the complex plane, to estimate $\tilde{x}$ from measurements of the controlled outputs y_i. The state observer of Eq.4.19 consists of a replica of the system of Eq.4.11 plus a linear gain multiplying the discrepancy between the actual and the estimated value of the output, and therefore, it is an extended Luenberger-type observer. It can be shown that the bounded stability of the closed-loop system resulting from the application of a robust output feedback controller (resulting from the combination of the controller of Eq.4.17 with the observer of Eq.4.19) to the particulate process model is guaranteed, provided that there exists a matrix L such that $C_L = \frac{1}{\mu}\bar{A}$ where μ is a sufficiently small positive parameter and $\bar{A}$ is a Hurwitz matrix. The reader may refer to the crystallization example of section 4 for an application of this approach for state estimation.

Remark 4.7: We note that the validity of the approach which we followed here to synthesize the nonlinear robust controller of Eq.4.17 relies on the large separation of slow and fast modes of the particulate process model of Eq.4.3. Despite the fact that the model of Eq.4.20 consists of a first-order hyperbolic PDE (population balance) coupled with a nonlinear integro-differential equation (solute mass balance), the approach followed here for the synthesis of robust feedback controllers is not applicable to hyperbolic PDE systems arising in the context of convection-reaction processes. More specifically, the dominant dynamic behavior of the system of Eq.4.1 is characterized by a small number of degrees of freedom (and thus, it can be described by low-order ODE systems which can be used for controller synthesis), while first-order hyperbolic PDE systems involve spatial differential operators whose eigenvalues cluster along vertical or nearly vertical asymptotes in the complex plane and thus, the controller synthesis problem has to be addressed directly on the basis of the hyperbolic PDE system (see Christofides and Daoutidis, 1998).

4. Application to a Continuous Crystallizer with Fines Trap

In this section, we apply the proposed robust control methodology to a continuous crystallizer with fines trap shown in Figure 4.1. The trap

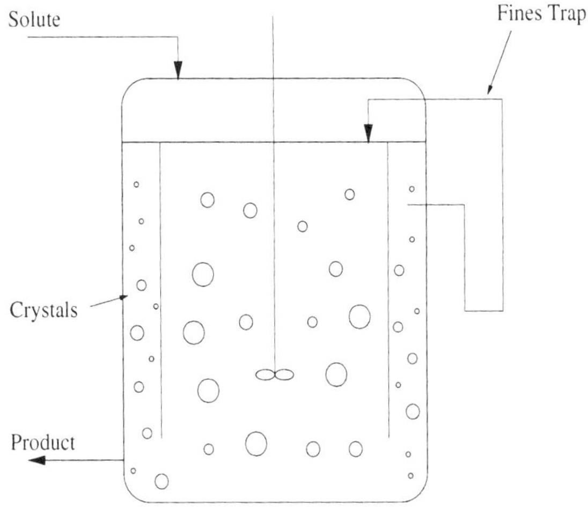

Figure 4.1. A continuous crystallizer with fines trap.

is used to remove small crystals and increase the mean crystal size. In a crystallizer, the precise regulation of the shape of the crystal size distribution is important because the CSD influences significantly the necessary liquid-solid separation and the properties of the product. Therefore, crystallization requires a population balance in order to be accurately described, analyzed and controlled. Under the standard assumptions of constant volume, mixed suspension, mixed product removal and nucleation of crystals of infinitesimal size, application of a population balance to the particulate phase and a mass balance to the continuous phase results in the following dynamic model for the crystallizer (Lei et al., 1971):

$$\begin{aligned} \frac{\partial n}{\partial \bar{t}} &= -\frac{\partial(R(\bar{t})n)}{\partial r} - \frac{n}{\tau} - \bar{h}(r)\frac{n}{\tilde{\tau}} + \delta(r-0)Q(\bar{t}) \\ \frac{dc}{d\bar{t}} &= \frac{(c_0-\rho)}{\bar{\epsilon}\tau} + \frac{(\rho-c)}{\tau} + \frac{(\rho-c)}{\bar{\epsilon}}\frac{d\bar{\epsilon}}{d\bar{t}} \end{aligned} \tag{4.20}$$

where $n(r,\bar{t})$ is the number of crystals of radius $r \in [0,\infty)$ at time $\bar{t}$ per unit volume of suspension, τ is the residence time, c is the solute concentration in the crystallizer, c_0 is the solute concentration in the feed, $\bar{\epsilon} = 1 - \int_0^{\infty} n(r,\bar{t})\frac{4}{3}\pi r^3 dr$ is the volume of liquid per unit volume of suspension, $R(\bar{t})$ is the growth rate, $\delta(r-0)$ is the standard Dirac function and $Q(\bar{t})$ is the nucleation rate. $1/\tilde{\tau} = F_0/V$ is the rate at which crystals are circulated through the fines trap (F_0 is the fines recirculation rate and V is the active volume of the crystallizer which is assumed to be constant) and $\bar{h}(r)$ expresses the desired selection curve for fines destruction (classification function). We assume that it is desirable to remove with the fines trap crystals of size r_m and smaller, and thus $\bar{h}(r)$ takes the form:

$$\bar{h}(r) = \begin{bmatrix} 1, & for \;\; r \leq r_m \\ 0, & for \;\; r > r_m \end{bmatrix} \tag{4.21}$$

In the population balance, the term $\delta(r-0)Q(\bar{t})$ accounts for the production of crystals of infinitesimal (zero) size via nucleation. $R(\bar{t})$ and $Q(\bar{t})$ are assumed to follow McCabe's law and Volmer's nucleation law, respectively:

$$R(\bar{t}) = k_1(c-c_s), \quad Q(\bar{t}) = \bar{\epsilon} k_2 e^{-\frac{k_3}{\left(\frac{c}{c_s}-1\right)^2}} \tag{4.22}$$

where k_1, k_2, k_3 are constants and c_s is the concentration of solute at saturation. Using the expressions for $Q(\bar{t})$ and $R(\bar{t})$, the system of Eq.4.20 can be written as:

$$\begin{aligned} \frac{\partial n}{\partial \bar{t}} &= -k_1(c-c_s)\frac{\partial n}{\partial r} - \frac{n}{\tau} - \bar{h}(r)\frac{n}{\tilde{\tau}} + \delta(r-0)\bar{\epsilon}k_2 e^{-\frac{k_3}{\left(\frac{c}{c_s}-1\right)^2}} \\ \frac{dc}{d\bar{t}} &= \frac{(c_0-\rho)}{\bar{\epsilon}\tau} + \frac{(\rho-c)}{\tau} + \frac{(\rho-c)}{\bar{\epsilon}}\frac{d\bar{\epsilon}}{d\bar{t}} \end{aligned} \tag{4.23}$$

The multivariable control problem is formulated as the one of controlling the crystal concentration and the solute concentration, i.e.:

$$\begin{aligned} y_1(\bar{t}) &= 8\pi\sigma^3 \int_0^\infty n(r,\bar{t})dr = \tilde{x}_0, \\ y_2(\bar{t}) &= \frac{(c(\bar{t})-c_s)}{(c_0-c_s)} = \tilde{y}(\bar{t}) \end{aligned} \tag{4.24}$$

by manipulating the flow rate of suspension through the fines trap and the inlet solute concentration, i.e.:

$$\begin{aligned} \bar{u}_1(\tilde{t}) &= \frac{1}{\tilde{\tau}} - \frac{1}{\tilde{\tau}_s}, \\ \bar{u}_2(\tilde{t}) &= \frac{c_0 - c_{0_s}}{c_0 - c_s} \end{aligned} \tag{4.25}$$

where $1/\tilde{\tau}_s$ and c_{0_s} denote the circulation rate of suspension in the fines trap and the inlet concentration at steady state, respectively. Since both manipulated variables directly enter the equations that describe the dynamics of the CSD and solute concentration, the crystallizer with the crystal concentration and solute concentration as controlled outputs and the flow rate of suspension through the fines trap and solute feed concentration as manipulated inputs is an approximately controllable system (see Semino and Ray, 1995a for a rigorous controllability analysis).

Uncertainties in the form of modeling errors in the pre-exponential factor of the nucleation rate, k_2, and the density of crystals, ρ, are introduced into the system. Specifically:

$$\begin{aligned} k_2 &= k_{2,nom} + 0.5k_{2,nom}sin(0.5t), \\ \rho &= \rho_{nom} + 0.1\rho_{nom} \end{aligned} \tag{4.26}$$

where $k_{2,nom}$ and ρ_{nom} represent the nominal values of the pre-exponential factor and the crystal density respectively. We also consider fast process dynamics arising from unmodeled dynamics of the controlled actuators and the measurement sensors. To account for the actuator dynamics, we consider the following dynamical system:

$$\begin{aligned} \epsilon_1 \dot{z}_1 &= -z_1 + z_2 \\ \epsilon_1 \dot{z}_2 &= -z_2 + \bar{u}_1 \\ \epsilon_2 \dot{z}_3 &= -z_3 + z_4 \\ \epsilon_2 \dot{z}_4 &= -z_4 + \bar{u}_2 \end{aligned} \tag{4.27}$$

where z_1 and z_3 are the values of $\bar{u}_1$ (flow rate of suspension through the fines trap) and $\bar{u}_2$ (inlet solute concentration) implemented on the process, and ϵ_1, ϵ_2 are small positive parameters. On the other hand, we assume that the sensor dynamics are described by the following dynamical system:

$$\begin{aligned} \epsilon_3 \dot{z}_5 &= -z_5 + z_6 \\ \epsilon_3 \dot{z}_6 &= -z_6 + y_1 \\ \epsilon_4 \dot{z}_7 &= -z_7 + z_8 \\ \epsilon_4 \dot{z}_8 &= -z_8 + y_2 \end{aligned} \tag{4.28}$$

where z_5 and z_7 are the values of y_1 (crystal size concentration) and y_2 (solute concentration) used in the controller, and ϵ_3, ϵ_4 are small positive parameters. The models for the actuator and sensor dynamics of Eqs.4.27-4.28 are coupled with the model of Eq.4.23 leading to the full model that describes the process.

Performing a two-time-scale decomposition to the full process model, one can show that its fast dynamics are stable, and that its slow dynamics are described by the ϵ-independent model of Eq.4.23 which will be used as the basis for controller synthesis. Owing to its distributed parameter nature, Eq.4.23 cannot be directly used for the synthesis of model-based feedback controllers. A model reduction procedure based on a combination of the method of moments and the approximation of the crystal size distribution with a Laguerre series expansion is used to reduce the system of Eq.4.23 into a small set of ordinary differential equations. This model reduction procedure is motivated from the fact that the dominant dynamics of the system of Eq.4.23 are characterized by a small number of degrees of freedom; see also chapter 2 for a detailed analysis of the dynamic behavior of a continuous crystallizer which reveals low-order dynamic behavior. Defining the νth moment of $\eta(r,\bar{t})$ as:

$$\mu_\nu = \int_0^\infty r^\nu n(r,\bar{t})dr, \quad \nu = 0, \ldots, \tag{4.29}$$

multiplying the population balance in Eq.4.23 by r^ν and integrating over all crystal sizes, the following infinite set of ordinary differential equations, which describes the rate of change of all the moments of the crystal size distribution and the solute concentration, is obtained:

$$\begin{aligned}
\frac{d\mu_0}{d\bar{t}} &= -\frac{\mu_0}{\tau} - \int_0^{r_m} \frac{n(r,\bar{t})}{\bar{\tau}} dr + \left(1 - \frac{4}{3}\pi\mu_3\right) k_2 e^{-\frac{k_3}{\left(\frac{c}{c_s}-1\right)^2}} \\
\frac{d\mu_\nu}{d\bar{t}} &= -\frac{\mu_\nu}{\tau} + \nu k_1 (c - c_s)\mu_{\nu-1} - \int_0^{r_m} r^\nu \frac{n(r,\bar{t})}{\bar{\tau}} dr, \quad \nu = 1, 2, 3, \ldots, \\
\frac{dc}{d\bar{t}} &= \frac{c_0 - c - 4\pi\tau(c - c_s)\mu_2(\rho - c)}{\tau\left(1 - \frac{4}{3}\pi\mu_3\right)}
\end{aligned} \tag{4.30}$$

Referring to the above system, note that it constitutes an unclosed set of moment equations owing to the nature of the classification function of the fines destruction. It is important to point out that when a fines trap is not used (i.e., $h(r) = 0$), then the first four moment equations and the concentration equation consist a closed set of differential equations that accurately describes the dominant dynamics of the crystallizer (see chapter 2 for details).

In order to close the set of moment equations, an approximate analytical expression for $n(r,\bar{t})$ in terms of the moments is needed in order to obtain a closed set of ordinary differential equations. Such an approximation using Laguerre polynomial expansion was suggested in Hulburt and Katz, 1964 and takes the following form:

$$n(r,\bar{t}) = \frac{\lambda}{a} p^{(\lambda)}\left(\frac{\lambda r}{a}\right) \sum_{n=0}^{\infty} \kappa_n L_n^{(\lambda)}\left(\frac{\lambda r}{a}\right) \tag{4.31}$$

where a and λ are functions of the moments of the crystal size distribution which are explicit functions of time. $L_n^{(\lambda)}$ are the nth order associated Laguerre polynomials:

$$L_n^{(\lambda)}(z) = \sum_{j=0}^{\infty} (-1)^j \frac{n!(n+\lambda-1)!}{j!(n-j)!(n+\lambda-1-j)!} z^{n-j} \qquad n = 0, 1, 2, \ldots \tag{4.32}$$

which are constructed by orthogonalizing the powers of z with respect to the Γ-distribution weighting function:

$$p^{(\lambda)}(z) = \frac{1}{(\lambda-1)!} z^{\lambda-1} e^{-z} \tag{4.33}$$

and κ_n takes the following form:

$$\kappa_n = \sum_{j=0}^{n} (-1)^j \frac{(\lambda-1)!}{j!(n+\lambda-1-j)!} \frac{(\lambda/a)^{n-j}}{(n-j)!} \mu_{n-j} \tag{4.34}$$

with the leading terms of κ_n being:

$$\begin{aligned} \kappa_0 &= \mu_0 \\ \kappa_1 &= \frac{1}{a}\mu_1 - \mu_0 \\ \kappa_2 &= \frac{1}{2a^2}\frac{\lambda}{\lambda+1}\mu_2 - \frac{1}{a}\mu_1 + \frac{1}{2}\mu_0 \end{aligned} \tag{4.35}$$

If we choose a and λ to be:

$$a = \frac{\mu_1}{\mu_0} \qquad \lambda = \frac{a^2}{\mu_2/\mu_0 - a^2} \tag{4.36}$$

so as to force κ_1 and κ_2 to be 0, Eq.4.31 becomes:

$$n(r,\bar{t}) = \frac{\lambda}{a} p^{(\lambda)}\left(\frac{\lambda r}{a}\right) \left[\mu_0 + \sum_{n=3}^{\infty} \kappa_n L_n^{(\lambda)}\left(\frac{\lambda r}{a}\right)\right] \tag{4.37}$$

Neglecting the terms in Eq.4.37 with $n = 3$ and higher results in the following approximation for the particle size distribution:

$$n(r,\bar{t}) = \frac{\lambda(\bar{t})}{a(\bar{t})} p^{(\lambda(\bar{t}))}\left(\frac{\lambda(\bar{t}) r}{a(\bar{t})}\right) \mu_0 \tag{4.38}$$

Substituting Eq.4.38 into Eq.4.30 and introducing the following set of dimensionless variables and parameters:

$$
\begin{gathered}
t = \frac{\bar{t}}{\tau}, \ \tilde{x}_0 = 8\pi\sigma^3\mu_0, \ \tilde{x}_1 = 8\pi\sigma^2\mu_1, \ \tilde{x}_2 = 4\pi\sigma\mu_2, \tilde{x}_3 = \frac{4}{3}\pi\mu_3, \ \ldots, \\
\sigma = k_1\tau(c_0 - c_s), Da = 8\pi\sigma^3 k_2\tau, F = \frac{k_3 c_s^2}{(c_0 - c_s)^2}, \ \alpha = \frac{(\rho - c_s)}{(c_0 - c_s)}, \\
\tilde{y} = \frac{(c - c_s)}{(c_0 - c_s)}
\end{gathered}
\tag{4.39}
$$

the following system is obtained:

$$
\begin{aligned}
\frac{d\tilde{x}_0}{dt} &= -\tilde{x}_0 + (1 - \tilde{x}_3)Dae^{\frac{-F}{\tilde{y}^2}} \\
&\quad -8\pi\sigma^3\mu_0\frac{\tau}{\tilde{\tau}}\frac{1}{(\lambda - 1)!}\left(\frac{\lambda}{a}\right)^{\lambda}\int_0^{r_m} r^{\lambda - 1}e^{-\frac{\lambda r}{a}}\,dr \\
\frac{d\tilde{x}_1}{dt} &= -\tilde{x}_1 + \tilde{y}\tilde{x}_0 - 8\pi\sigma^2\mu_0\frac{\tau}{\tilde{\tau}}\frac{1}{(\lambda - 1)!}\left(\frac{\lambda}{a}\right)^{\lambda}\int_0^{r_m} r^{\lambda}e^{-\frac{\lambda r}{a}}\,dr \\
\frac{d\tilde{x}_2}{dt} &= -\tilde{x}_2 + \tilde{y}\tilde{x}_1 - 4\pi\sigma\mu_0\frac{\tau}{\tilde{\tau}}\frac{1}{(\lambda - 1)!}\left(\frac{\lambda}{a}\right)^{\lambda}\int_0^{r_m} r^{\lambda + 1}e^{-\frac{\lambda r}{a}}\,dr \\
\frac{d\tilde{x}_3}{dt} &= -\tilde{x}_3 + \tilde{y}\tilde{x}_2 - \frac{4}{3}\pi\mu_0\frac{\tau}{\tilde{\tau}}\frac{1}{(\lambda - 1)!}\left(\frac{\lambda}{a}\right)^{\lambda}\int_0^{r_m} r^{\lambda + 2}e^{-\frac{\lambda r}{a}}\,dr \\
&\ \vdots \\
\frac{d\tilde{y}}{dt} &= \frac{1 - \tilde{y} - (\alpha - \tilde{y})\tilde{y}\tilde{x}_2}{1 - \tilde{x}_3}
\end{aligned}
\tag{4.40}
$$

On the basis of the system of Eq.4.40, it is clear that the moments of order four and higher do not affect those of order three and lower, which means that the dominant dynamics of the system of Eq.4.30 can be adequately captured by the first four moment equations and the solute mass balance equation. The expressions of the corresponding dimensionless parameters can be obtained by substituting Eq.4.26 into Eq.4.39 to yield the following:

$$
\begin{aligned}
Da &= Da_{nom} + 0.5Da_{nom}sin(0.5t), \\
\alpha &= \alpha_{nom} + \frac{0.1\rho_{nom}}{c_0 - c_s}
\end{aligned}
\tag{4.41}
$$

where $Da_{nom} = 8\pi\sigma^3 k_{2,nom}\tau$ and $\alpha_{nom} = \dfrac{(\rho_{nom} - c_s)}{(c_0 - c_s)}$. Substituting the formulation of the control problem of Eqs.4.24 and 4.25 into the fifth-order moment model, we obtain a system of the following general form:

$$\begin{aligned} \frac{d\tilde{x}}{dt} &= \tilde{f}(\tilde{x}) + \tilde{g}_1(\tilde{x})\bar{u}_1 + \tilde{g}_2(\tilde{x})\bar{u}_2 + \tilde{w}_1(\tilde{x})\theta_1 + \tilde{w}_2(\tilde{x})\theta_2 \\ y_1 &= \tilde{h}_1(\tilde{x}), \qquad y_2 = \tilde{h}_2(\tilde{x}) \end{aligned} \tag{4.42}$$

where $\tilde{x}$ denotes the vector $[\tilde{x}_0\ \tilde{x}_1\ \tilde{x}_2\ \tilde{x}_3\ \tilde{y}]^T$, $\tilde{f}(\tilde{x})$, $\tilde{g}_1(\tilde{x})$ and $\tilde{g}_2(\tilde{x})$ are vector functions and $\tilde{h}_1(\tilde{x})$ and $\tilde{h}_2(\tilde{x})$ are scalar functions whose explicit form is omitted for brevity.

Utilizing the formula of Theorem 4.1, we synthesize a robust nonlinear controller on the basis of the model of Eq.4.42 which has the following form:

$$\begin{aligned} \bar{u}_1(t) &= [\beta_{12} L_{\tilde{g}_1} \tilde{h}(\omega)]^{-1} \{v_1 - \beta_{10} y_1 - \beta_{11} L_{\tilde{f}} \tilde{h}_1(\omega) \\ &\quad -\chi_1 \frac{y_1 - v_1}{|y_1 - v_1| + \phi_1} 0.5 Da_{nom} |(1 - \omega_3) e^{\frac{-F}{\omega_4{}^2}}|\} \\ \bar{u}_2(t) &= [\beta_{22} L_{\tilde{g}_2} \tilde{h}(\omega)]^{-1} \{v_2 - \beta_{20} y_1 - \beta_{21} L_{\tilde{f}} \tilde{h}_2(\omega) \\ &\quad -\chi_2 \frac{y_2 - v_2}{|y_2 - v_2| + \phi_2} \frac{0.1\rho_{nom}}{(c_0 - c_s)} |\frac{\omega_4 \omega_2}{1 - \omega_3}|\} \end{aligned} \tag{4.43}$$

where v_1 and v_2 are the set-points for the two outputs, and $\beta_{10}, \beta_{11}, \beta_{12}, \beta_{20}, \beta_{21}, \beta_{22}, Da_{nom}, \rho_{nom}, \phi_1, \phi_2, \chi_1, \chi_2$ are parameters which are given in Tables 4.1 and 4.2. The practical implementation of the robust

σ	=	$k_1\tau(c_0 - c_s)$	=	1.0	mm
Da_{nom}	=	$8\pi\sigma^3 k_{2,nom}\tau$	=	200.0	
F	=	$k_3 c_s^2/(c_0 - c_s)^2$	=	3.0	
α_{nom}	=	$(\rho_{nom} - c_s)/(c_0 - c_s)$	=	40.0	

Table 4.1. Dimensionless parameters.

nonlinear controller of Eq.4.43 is achieved by employing the following

β_{10}	=	1
β_{11}	=	1.5
β_{12}	=	1.5
β_{20}	=	1
β_{21}	=	1.5
β_{22}	=	1.5
L_1	=	$[1\ 0\ 0\ 0\ 0]^T$
L_2	=	$[0\ 0\ 0\ 0\ 1]^T$
ϕ_1	=	0.0022
ϕ_2	=	0.009
χ_1	=	1.53
χ_2	=	5.48

Table 4.2. Controller parameters.

nonlinear state observer:

$$
\begin{aligned}
\frac{d\omega_0}{dt} &= -\omega_0 + (1-\omega_3)Dae^{\frac{-F}{\tilde{y}^2}} \\
&\quad -\bar{u}_1(t)8\pi\sigma^3\mu_0\tau\frac{1}{(\lambda-1)!}\left(\frac{\lambda}{a}\right)^\lambda\int_0^{r_m} r^{\lambda-1}e^{-\frac{\lambda r}{a}}dr \\
&\quad +L_{10}(\tilde{h}_1(\tilde{x})-\tilde{h}_1(\omega))+L_{20}(\tilde{h}_2(\tilde{x})-\tilde{h}_2(\omega)) \\
\frac{d\omega_1}{dt} &= -\omega_1+\tilde{y}\omega_0-\bar{u}_1(t)8\pi\sigma^2\mu_0\tau\frac{1}{(\lambda-1)!}\left(\frac{\lambda}{a}\right)^\lambda\int_0^{r_m} r^{\lambda}e^{-\frac{\lambda r}{a}}dr \\
&\quad +L_{11}(\tilde{h}_1(\tilde{x})-\tilde{h}_1(\omega))+L_{21}(\tilde{h}_2(\tilde{x})-\tilde{h}_2(\omega)) \\
\frac{d\omega_2}{dt} &= -\omega_2+\tilde{y}\omega_1-\bar{u}_1(t)4\pi\sigma\mu_0\tau\frac{1}{(\lambda-1)!}\left(\frac{\lambda}{a}\right)^\lambda\int_0^{r_m} r^{\lambda+1}e^{-\frac{\lambda r}{a}}dr \\
&\quad +L_{12}(\tilde{h}_1(\tilde{x})-\tilde{h}_1(\omega))+L_{22}(\tilde{h}_2(\tilde{x})-\tilde{h}_2(\omega)) \\
\frac{d\omega_3}{dt} &= -\omega_3+\tilde{y}\omega_2-\bar{u}_1(t)\frac{4}{3}\pi\mu_0\tau\frac{1}{(\lambda-1)!}\left(\frac{\lambda}{a}\right)^\lambda\int_0^{r_m} r^{\lambda+2}e^{-\frac{\lambda r}{a}}dr \\
&\quad +L_{13}(\tilde{h}_1(\tilde{x})-\tilde{h}_1(\omega))+L_{23}(\tilde{h}_2(\tilde{x})-\tilde{h}_2(\omega)) \\
\frac{d\omega_4}{dt} &= \frac{1-\omega_4-(\alpha-\omega_4)\omega_4\omega_2}{1-\omega_3}+\frac{1}{1-\tilde{\omega}_3}\bar{u}_2(t)+L_{14}(\tilde{h}_1(\tilde{x})-\tilde{h}_1(\omega)) \\
&\quad +L_{24}(\tilde{h}_2(\tilde{x})-\tilde{h}_2(\omega))
\end{aligned} \tag{4.44}
$$

where ω denotes the observer state vector whose dimension is equal to that of $\tilde{x}$, and $\tilde{h}_1(\tilde{x})$ and $\tilde{h}_2(\tilde{x})$ are the measured outputs. The state observer of Eq.4.44 consists of a replica of the system of Eq.4.42 plus two linear gain vectors $L_1 = [L_{10}\ L_{11}\ L_{12}\ L_{13}\ L_{14}]$ and $L_2 = [L_{20}\ L_{21}\ L_{22}\ L_{23}\ L_{24}]$ multiplying the discrepancy between the actual and the estimated values of the outputs. The values of the gains L_1 and L_2 are given in Table 4.3. The practical implementation of the nonlinear robust output feedback controller of Eqs.4.43-4.44 requires on-line measurements of the controlled outputs, $\tilde{x}_0$ (crystal concentration) and $\tilde{y}$ (solute concentration); in practice, measurements of $\tilde{x}_0$ can be obtained by using, for example, light scattering (see Bohren and Huffman, 1983; Rawlings et al., 1993 for details), and measurements of $\tilde{y}$ (solute concentraton) can be obtained by using a mass spectrometer.

The performance of the nonlinear robust output feedback controller of Eqs.4.43-4.44 was tested through numerical simulations. The values of the system parameters and the ones in their corresponding dimensionless forms (Eq.4.39) are shown in Tables 4.3 and 4.1 respectively. In all

c_0	=	1000.0	$kg\ m^{-3}$
c_s	=	980.2	$kg\ m^{-3}$
ρ_{nom}	=	1770.0	$kg\ m^{-3}$
τ	=	1.0	hr
$1/\tilde{\tau}_s$	=	0.0	hr^{-1}
k_1	=	5.065×10^{-2}	$mm\ m^3\ kg^{-1}\ hr^{-1}$
$k_{2,nom}$	=	7.958	$mm^{-3}\ hr^{-1}$
k_3	=	1.217×10^{-3}	
r_m	=	1	mm

Table 4.3. Process parameters.

simulation runs, the following initial condition:

$$\begin{aligned} n(r,0) &= (2.189 \times 10^3)e^{-1.168r}mm^{-4}, \\ c(0) &= 992.1\ kg\ m^{-3} \end{aligned} \tag{4.45}$$

was used for the process model of Eq.4.23 and the finite difference method with 1000 discretization points was used for the simulations. The initial conditions for the observer states were computed numerically by using the initial conditions of Eq.4.45. We note that even though the number of discretization points, 1000, used to solve the system of Eq.4.20 is very large (owing to the poor convergence properties of the finite-difference scheme), the computation of an accurate (i.e., independent of the discretization) solution is critical for the thorough evaluation of the performance of a nonlinear feedback controller synthesized on the basis of a low-order approximation of the distributed parameter system of Eq.4.20. The adequacy of 1000 discretization points to yield an accurate solution was established through extensive simulations of the open- and closed-loop systems.

In the first set of simulation runs, we implemented the nonlinear robust multivariable controller on the crystallizer model of Eq.4.23 with the actuator dynamics of Eq.4.27 in the presence of parametric uncertainty in the pre-exponential factor of the nucleation rate, k_2, and the density of crystals, ρ. A 0.03 decrease in the value of the set-point from the initial conditions was applied to the first controlled output, $\tilde{x}_0$ ($v_1 = 0.015$), while the set-point for the second output, $\tilde{y}$, was set to be the same as the initial conditions ($v_2 = 0.5996$). The closed-loop output profiles are shown in Figure 4.2 (solid lines) with $\epsilon_1 = 0.1$ for the first manipulated input (flowrate of suspension through fines trap) and $\epsilon_2 = 0.02$ for the second manipulated input (inlet solute concentration); these are the largest values for ϵ_1, ϵ_2 for which an acceptable closed-loop output response is achieved with stability. Clearly the controller regulates the outputs to the set-point values attenuating the effect of the time-varying uncertainty on the process outputs and being robust to unmodeled actuator dynamics. Figure 4.3 shows the profiles of the manipulated inputs (solid lines). It is observed that the control action computed by the robust controller exhibits an oscillatory behavior in order to compensate for the time-variation of the parameter Da. For the sake of comparison, we also implemented the nonlinear multivariable controller of Eq.4.43 with $\chi = 0$ (i.e., no uncertainty compensation is included in the controller). Figure 4.2 displays the closed-loop output profiles (dashed lines) and Figure 4.3 displays the corresponding manipulated input profiles (dashed lines). Clearly, this controller cannot compensate for the effect of the time-varying uncertainty leading to very poor closed-loop performance. Finally, to evaluate the ability of the controller in attenuating the effect of the time-varying uncertain variables in the entire CSD, the closed-loop profile of the evolution of the CSD under robust nonlinear control is shown in Figure 4.4 (bottom plot) and compared to the one of the open-loop process (top plot). It is obvious that fluctuations in the CSD are effectively damped by the use of robust nonlinear control.

In the second set of simulation runs, we implemented the nonlinear robust multivariable controller on the crystallizer model of Eq.4.23 with the actuator dynamics of Eq.4.27 and the sensor dynamics of Eq.4.28

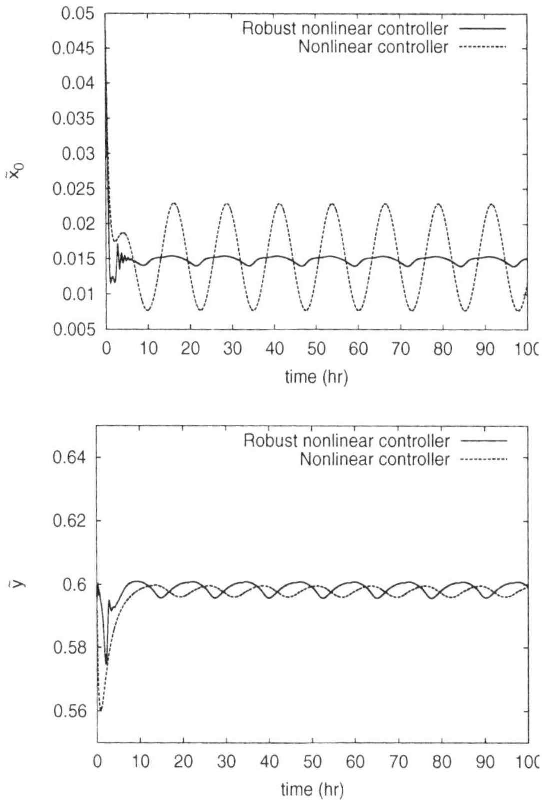

Figure 4.2. Closed-loop output profiles for $\tilde{x}_0$ (top plot) and $\tilde{y}$ (bottom plot) under robust nonlinear controller (solid line) and nonlinear controller which does not account for uncertainty (dashed line) - actuator dynamics are included in the process model.

in the presence of the same parametric uncertainty considered in the first set of simulation runs. Again, a 0.03 decrease in the value of the set-point from the initial conditions was applied to the first controlled output, $\tilde{x}_0$ ($v_1 = 0.015$), while the set-point for the second controlled output, $\tilde{y}$, was set to be $v_2 = 0.5996$. The closed-loop output profiles are shown in Figure 4.5 (solid lines) with $\epsilon_1 = 0.1$ for the first manipulated input (flowrate of suspension through fines trap), $\epsilon_2 = 0.02$ for the second manipulated input (inlet solute concentration), $\epsilon_3 = 0.01$ for the first measured output (crystal concentration), $\epsilon_4 = 0.01$ for the second measured output (solute concentration); these are the largest values for $\epsilon_1, \epsilon_2, \epsilon_3, \epsilon_4$ for which an acceptable closed-loop output response is achieved with stability. Clearly the controller regulates the outputs

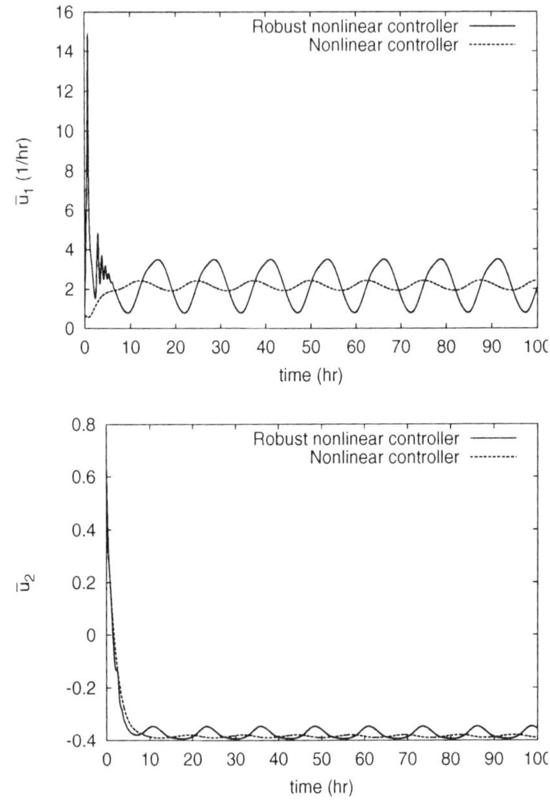

Figure 4.3. Manipulated input profiles for $1/\tilde{\tau}$ (top plot) and c_0 (bottom plot) under robust nonlinear controller (solid line) and nonlinear controller which does not account for uncertainty (dashed line) - actuator dynamics are included in the process model.

to the set-point values minimizing the effect of the time-varying uncertainty. We also implemented the nonlinear multivariable controller of Eq.4.43 with $\chi = 0$ (i.e., no uncertainty compensation is included in the controller). Figure 4.5 shows the closed-loop output profiles (dashed lines). It is clear that this controller cannot attenuate the effect of the uncertainty on the outputs leading to poor performance. Figure 4.6 shows the profiles of the manipulated inputs of the robust controller (solid lines) and the ones of the nonlinear controller which does not compensate for uncertainty (dashed lines). Again, the control actions computed by the robust controller exhibit oscillatory behavior to compensate for the time-varying uncertainty. Finally, the ability of the robust nonlinear controller to attenuate the uncertainty can be also seen in Figure 4.7

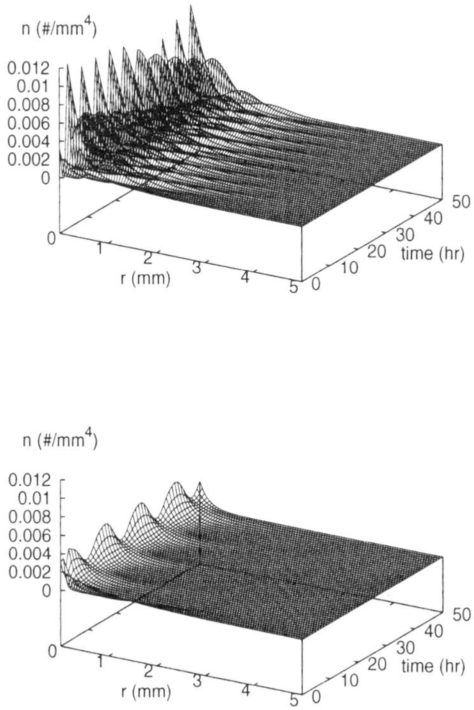

Figure 4.4. Profile of evolution of crystal size distribution, open-loop (top plot) and closed-loop under robust nonlinear controller (bottom plot) - actuator dynamics are included in the process model.

(bottom plot) where the evolution of the CSD under robust nonlinear control is shown and compared with the open-loop profile (top plot).

Remark 4.8: It is important to note that we have also performed simulations of the process under proportional-integral control and obtained a very poor closed-loop performance (worst than the performance obtained under nonlinear control without uncertainty compensation). This is expected because proportional-integral control cannot effectively deal with the presence of time-varying uncertainty and significant nonlinearities in the process model.

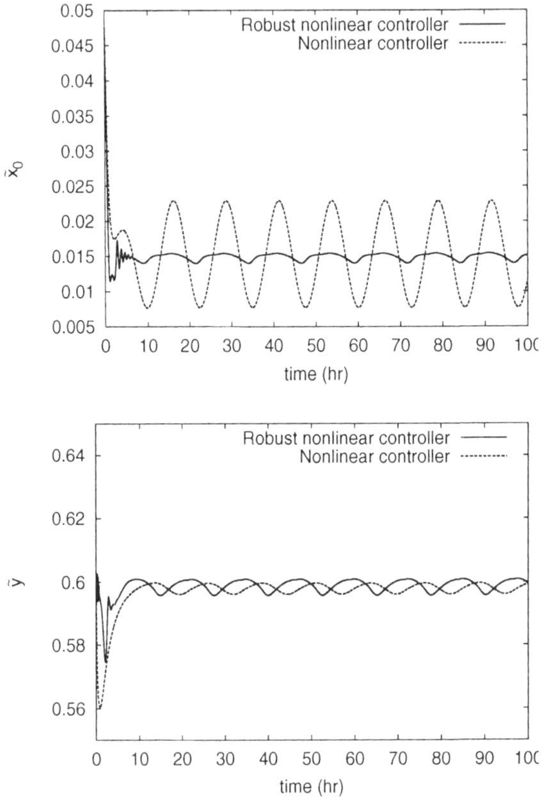

Figure 4.5. Closed-loop output profiles for $\tilde{x}_0$ (top plot) and $\tilde{y}$ (bottom plot) under robust nonlinear controller (solid line) and nonlinear controller which does not account for uncertainty (dashed line) - actuator and sensor dynamics are included in the process model.

5. Conclusions

A general method was presented for the synthesis of robust nonlinear controllers for a broad class of particulate processes described by population balances that explicitly handle time-varying uncertain variables (e.g., unknown process parameters and disturbances) and unmodeled dynamics (e.g., fast actuator and sensor dynamics not included in the process model). The robust controllers are synthesized on the basis of finite-dimensional approximations of uncertain population balances via Lyapunov's direct method and enforce stability in the closed-loop system, attenuation of the effect of uncertain variables and achieve particle size distributions with desired characteristics. The robustness of the pro-

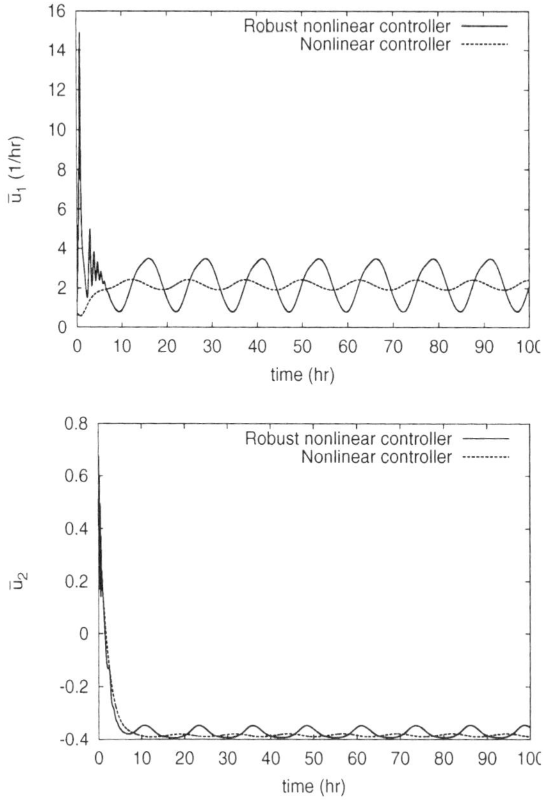

Figure 4.6. Manipulated input profiles for $1/\tilde{\tau}$ (top plot) and c_0 (bottom plot) under robust nonlinear controller (solid line) and nonlinear controller which does not account for uncertainty (dashed line) - actuator and sensor dynamics are included in the process model.

posed controllers with respect to stable and sufficiently fast unmodeled dynamics was established through a singular perturbation analysis. The controllers were applied to a continuous crystallizer with fines trap in which the nucleation rate and the crystal density change with time and the actuator and sensor dynamics are not included in the model used for the synthesis of the controller. Simulation runs of the closed-loop system clearly demonstrated the ability of the controllers to attenuate the uncertainty and achieve a crystal size distribution with desired characteristics, and documented their superiority over nonlinear controllers that do not account for the presence of uncertainty.

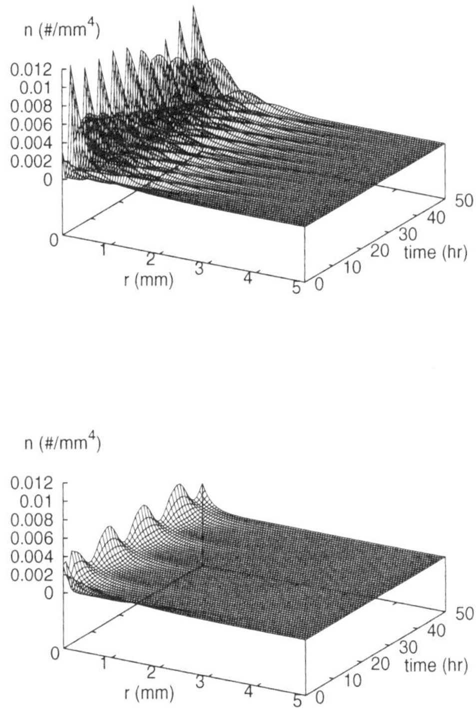

Figure 4.7. Profile of evolution of crystal size distribution, open-loop (top plot) and closed-loop under robust nonlinear controller (bottom plot) - actuator and sensor dynamics are included in the process model.

Chapter 5

CONSTRAINED CONTROL OF PARTICULATE PROCESSES

1. Introduction

This chapter focuses on the development of a general methodology for the analysis and control of constrained spatially-homogeneous particulate processes modeled by population balance equations and its application to a continuous crystallizer. Initially, the method of weighted residuals is used for the construction of finite-dimensional ODE systems that accurately reproduce the dominant dynamics of the particulate process. These ODE systems are then used to analyze the limitations imposed by input constraints on the ability to modify the dynamics of the particulate process, leading to an explicit characterization of the set of admissible set-points that can be achieved in the presence of constraints. This information together with the derived ODE systems are then used as the basis for the synthesis of practically-implementable nonlinear bounded output feedback controllers that enforce exponential stability in the closed-loop system and achieve particle size distributions with desired characteristics in the presence of active input constraints. Precise closed-loop stability conditions are given and controller implementation issues are addressed. The proposed methodology is applied to a continuous crystallizer. The results of this chapter were first presented in El-Farra et al., 2001.

2. Preliminaries

2.1 Particulate Process Model with Input Constraints

We focus on spatially homogeneous (well-mixed) particulate processes with simultaneous particle growth, nucleation, agglomeration and breakage and consider the case of a single internal particle coordinate, which is assumed to be the particle size. Applying a population balance to the particle phase as well as material and energy balances to the continuous phase, we obtain the following general nonlinear system of partial integro-differential equations:

$$\begin{aligned} \frac{\partial n}{\partial t} &= -\frac{\partial(G(x,r)n)}{\partial r} + w(n,x,r), \;\; n(0,t) = b(x(t)) \\ \dot{x} &= f(x) + g(x)sat(u(t)) + A\int_0^{r_{max}} a(n,r,x)dr \end{aligned} \tag{5.1}$$

where $n(r,t) \in [L_2[0,r_{max}), \mathbb{R}]$ is the size distribution function which is assumed to be a continuous and sufficiently smooth function of its arguments (we use the symbol $L_2[0,r_{max})$ to denote a Hilbert space of continuous functions defined on the interval $[0,r_{max})$), $r \in [0,r_{max})$ is the particle size (r_{max} is the maximum particle size, which may be infinity), t is the time, $x \in \mathbb{R}^n$ is the vector of state variables which describe properties of the continuous phase (e.g., solute concentration, temperature and pH in a crystallizer), $u(t) = [u_1(t)\; u_2(t)\; \cdots\; u_m(t)]^T \in \mathbb{R}^m$ is the vector of manipulated inputs, and sat refers to the saturation function defined by:

$$sat(u_i) = \left\{ \begin{array}{lcc} u_{i,min} & if & u_i < u_{i,min} \\ u_i & if & u_{i,min} \leq u_i \leq u_{i,max} \\ u_{i,max} & if & u_i > u_{i,max} \end{array} \right\} \tag{5.2}$$

where $u_i \in \mathbb{R}$ and for a vector $u \in \mathbb{R}^m$,
$sat(u) = [sat(u_1)\; sat(u_2)\; \cdots\; sat(u_m)]^T$. The presence of the sat operator in Eq.5.1 signifies the presence of hard constraints on the manipulated input. Such constraints may arise naturally due to inherent limitations on the capacity of control actuators used to regulate particulate processes or may be imposed for economic or safety reasons.

Following Chapters 2 and 4, we define a vector of controlled outputs to express the various control objectives (e.g., regulation of total number of particles, mean particle size, temperature, pH, etc.) as follows:

$$y_i(t) = h_i(\int_0^{r_{max}} c_\kappa(r)n(r,t)dr, x), \;\; i = 1,\ldots,m, \;\; \kappa = 1,\ldots,l \tag{5.3}$$

where $y_i(t)$ is the ith controlled output, $h_i(\int_0^{r_{max}} c_\kappa(r)n(r,t)dr, x)$ is a nonlinear scalar smooth function of its arguments and $c_\kappa(r)$ is a known smooth function of r which depends on the desired performance specifications. To simplify the notation in our theoretical development, we assume the availability of on-line measurements of the controlled output $y_i(t)$. Finally, to state the theoretical results of this chapter, we will need the definition of a class KL function. In particular, a function $\beta(s,t)$ is said to be of class KL if, for each fixed t, the function $\beta(s,\cdot)$ is continuous, increasing, and zero at zero and, for each fixed s, the function $\beta(\cdot,t)$ is nonincreasing and tends to zero at infinity.

2.2 Motivating Example: A Continuous Crystallizer

Input constraints restrict our ability to modify the dynamics of the process in the desired manner. In order to motivate our discussion on the problem of input constraints and the approach that we propose for dealing with this problem (detailed in the next section), we highlight first in this section, by means of the continuous crystallizer example introduced in chapter 2, some of the key issues that need to be addressed for the effective control of particulate processes with input constraints. In particular, we demonstrate some of the fundamental limitations imposed by input constraints on our ability to steer particulate process dynamics as well as some of the detrimental effects of input constraints on the performance of conventional feedback controller designs. The mathematical model for this crystallizer is given by:

$$\begin{aligned}\frac{\partial n}{\partial \bar{t}} &= -k_1(c-c_s)\frac{\partial n}{\partial r} - \frac{n}{\tau} + \delta(r-0)\bar{\epsilon}k_2 e^{-\frac{k_3}{\left(\frac{c}{c_s}-1\right)^2}} \\ \frac{dc}{d\bar{t}} &= \frac{(c_0-\rho)}{\bar{\epsilon}\tau} + \frac{(\rho-c)}{\tau} + \frac{(\rho-c)}{\bar{\epsilon}}\frac{d\bar{\epsilon}}{d\bar{t}}\end{aligned} \tag{5.4}$$

where $n(r,\bar{t})$ is the number of crystals of radius $r \in [0,\infty)$ at time $\bar{t}$ per unit volume of suspension, τ is the residence time, c is the solute concentration in the crystallizer, c_0 is the solute concentration in the feed, $\bar{\epsilon} = 1 - \int_0^{\infty} n(r,\bar{t})\frac{4}{3}\pi r^3 dr$ is the volume of liquid per unit volume of suspension, c_s is the concentration of solute at saturation, k_1, k_2, and k_3 are constants, and $\delta(r-0)$ is the standard Dirac function. The term containing the Dirac function in Eq.5.4 accounts for the production of crystals of infinitesimal (zero) size via nucleation. The parameters used for this crystallizer process model are given in Table 2.1. It was shown in Chapter 2 that this crystallizer exhibits highly oscillatory behavior resulting from the interplay between the growth and nucleation terms in the population balance (unstable steady-state surrounded by a stable limit cycle.)

The control objective is to stabilize the crystallizer and achieve a crystal-size distribution with desired mass by manipulating the solute feed concentration. The manipulated input is therefore taken to be $u(t) = \dfrac{c_0 - c_{0s}}{c_0 - c_s}$, where c_{0s} is the steady-state solute feed concentration, and the controlled output is defined as

$$y(\bar{t}) = 8\pi\sigma^3 \int_0^\infty n(r,\bar{t})dr = \tilde{x}_0 \tag{5.5}$$

where $\sigma = k_1\tau(c_0 - c_s)$. Note that both are in dimensionless form. To achieve the desired control objective, we implement a conventional proportional-integral (PI) control scheme

$$\begin{aligned} \frac{d\eta}{dt} &= e \\ u &= K_c[e + \frac{1}{\tau_I}\eta] \end{aligned} \tag{5.6}$$

on the crystallizer process model of Eq.5.4, where $e = y_{sp} - y$ and u is the controller output. If there are no limits on the maximum solute feed concentration that can be used to achieve the desired control objective, then the controller output and process input are the same. However, suppose that there are limitations on the maximum solute feed concentration, i.e, $|u| \leq u_{max}$. In this case, if the solute feed concentration calculated by the controller is, for example, higher than u_{max}, then the actual solute concentration fed to the crystallizer will be just u_{max} leading to a mismatch between the controller output and actual crystallizer input.

Initially, closed-loop simulation runs were performed to test the ability of the PI controller to achieve the desired control objective in the absence of limitations (constraints) on the maximum solute feed concentration. Figure 5.1 depicts the dimensionless crystal concentration ($\tilde{x}_0$) and dimensionless solute feed concentration (u) profiles for a 0.4 increase in the set-point in the absence of any constraints on the manipulated input (i.e., $|u| < \infty$.) It is evident from the figure that the PI controller manages eventually to drive the output to its new set-point. By contrast, Figure 5.2 depicts the closed-loop response to the same 0.4 increase in the set-point in the presence of input constraints. The solid line shows the response when $u \in [0, 2]$ and the dashed line shows the response when $u \in [0, 6]$. As can be seen from this figure, the PI controller fails (regardless of the PI tuning parameters) to stabilize the crystallizer when the input is constrained in the interval $[0, 2]$, leading to sustained oscillations, while it successfully drives the crystallizer output (despite the poor transient performance) to the desired set-point when the input is constrained in the interval $[0, 6]$.

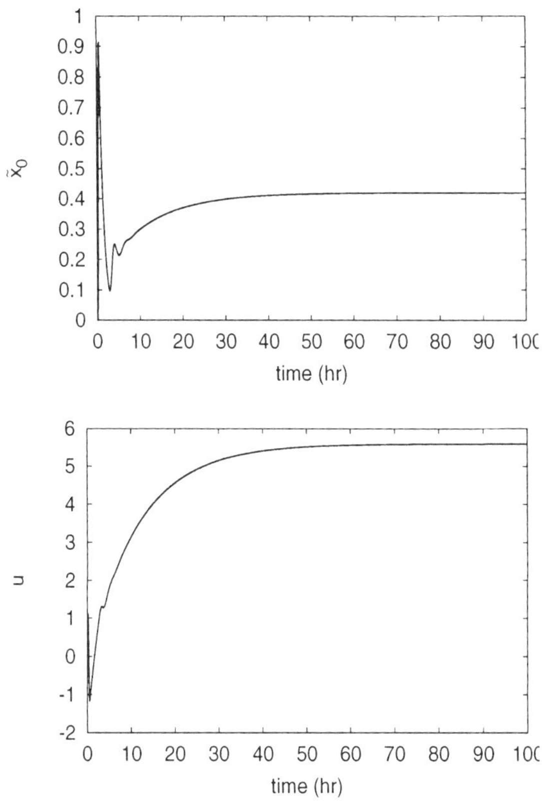

Figure 5.1. Controlled output (crystal concentration) and manipulated input (solute concentration) profiles under PI control in the absence of input constraints.

The contrast between the two different responses in Figure 5.2 illustrates an example of how input constraints place fundamental limitations on our ability to steer the particulate process to a desired set-point under any control law (linear or nonlinear). To understand these limitations, we refer to the input profiles in Figure 5.2 which show that in order for the process output to reach a set-point of 0.4, the process input must reach a steady state value of 5.3, i.e. outside $[0, 2]$. Therefore, when the input is constrained in $[0, 2]$, the process output cannot accommodate the requested set-point change (regardless of the controller used) owing to the fundamental inadmissibility of the desired set-point when

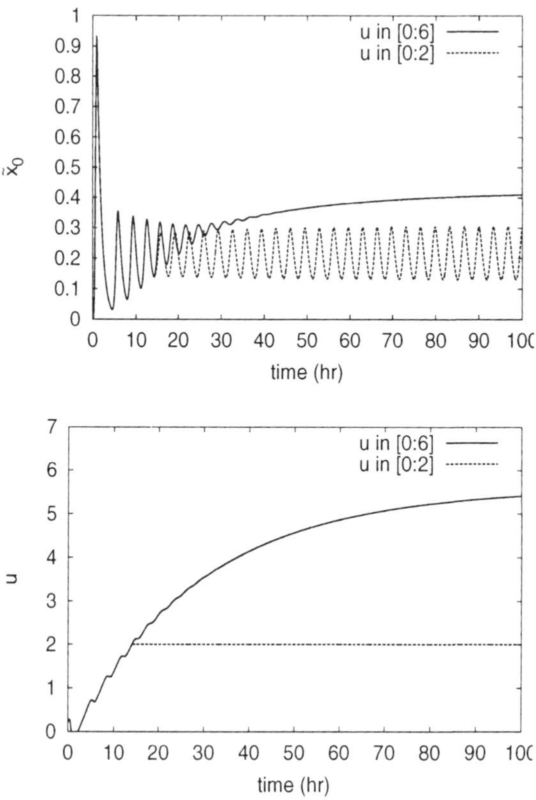

Figure 5.2. Controlled output (crystal concentration) and manipulated input (solute concentration) profiles under PI control when $u \in [0,2]$ (dashed lines) and when $u \in [0,6]$ (solid lines).

$u \in [0,2]$. Relaxing the constraints to $[0,6]$ consequently allows the process input to reach a steady state value of 5.3 and steer the process output to a value of 0.4. It is important to keep in mind therefore that this problem of set-point inadmissibility is independent of the controller used and cannot be remedied simply by choosing another controller in place of the PI controller.

The above reasoning immediately raises the following important question: given our knowledge of the constraints on the process input, how can we identify a priori (before controller implementation and without running closed-loop simulations) whether a given set-point is admissible? It is clear from Figure 5.2 that lack of this information can easily lead to a poor (i.e. inadmissible) choice of the set-point and subsequent instability. In the absence of this information, it may not even be clear what the underlying cause of instability is. This, in turn, may prompt the control engineer to suspect the particular controller used and consequently waste time and effort in re-tuning the controller or trying other control schemes without realizing that the problem is one that cannot be solved by any controller and is rather due to the inherent limitations imposed by the constraints. These considerations motivate the need for a systematic approach that explicitly characterizes the limitations imposed by input constraints on our ability to steer the particulate process in a desired direction. Such an approach is needed to provide process operators with the necessary knowledge of which set-point changes are feasible and can be achieved in the presence of known constraints on the manipulated input, *independent of the specific control scheme to be used.* The a priori availability of this kind of feasibility information is a pre-requisite for any effective controller design method that can successfully address the problems caused by input constraints on the operation of the process.

Another important issue that the above crystallizer example highlights is that of the particular choice of the controller to achieve the desired control objective, once that objective is determined to be admissible. Recall from Figure 5.2 (dashed line) that although the set-point is admissible when $u \in [0, 6]$ and that the PI controller stabilizes the process there, the transient performance is poor (compare with Figure 5.1). This performance deterioration is a direct consequence of input saturation for about two hours in the beginning of crystallizer operation leading to a more sluggish and oscillatory response. While it is possible to reduce the response settling time by further re-tuning of the PI controller, this comes at the expense of unreasonably large overshoot in the process output. It is important to point out here that owing to the complex nonlinear dynamics of the crystallizer, the presence of input constraints, and the large set-point change considered, there are no systematic guidelines for tuning the PI controller. Instead, the PI controller was tuned through extensive trial and error to obtain the best possible performance that yields reasonable overshoot (compared to that of the nonlinear controller shown later in Figure 5.6).

The "traditional" approaches of designing/tuning the PI controller on the basis of a linearized model of the process are inadequate because they do not account for either the strong nonlinearities present or the input constraints. The performance of any controller designed in this way deteriorates, even in the absence of constraints, as the process moves further away from the steady state around which the model was linearized. In addition to the performance problem, conventional control schemes (such as the PI controller) which are not designed to explicitly handle the presence of input constraints, lack any a priori guarantees regarding stability of the constrained closed-loop system starting from a given initial condition. *It is well known that even if a set-point is admissible, not every initial condition can be used to reach it.* This initial condition must belong to the region of closed-loop stability. Most conventional control schemes (linear and nonlinear) however do not provide any systematic way of estimating this region or identifying initial conditions starting from where closed-loop stability is guaranteed in the presence of input constraints. All these considerations combined clearly motivate the need for an alternative, effective and direct control strategy that can handle the presence of input constraints explicitly in the controller design and provide, simultaneously, an explicit characterization of the region of guaranteed closed-loop stability starting from where the requested stability and performance of the particulate process can be guaranteed in the presence of input constraints.

Motivated by the above discussion, the development of a rigorous, yet practical, framework for the analysis and control of constrained particulate processes is the subject of this paper. We commence in the next section with an outline of the proposed framework, which will serve as the road map for our development throughout this chapter.

3. Methodological Framework for Analysis and Control

Owing to its distributed parameter nature, the system of Eq.5.1 cannot be used directly as the basis for either the analysis of the limitations imposed by input constraints on particulate process dynamics or the synthesis of practically implementable (low-order) nonlinear controllers that cope effectively with the problem of constraints. This fact, together with the realization that the dominant dynamics of particulate processes are characterized by a small number of degrees of freedom, motivate employing the following methodology for the analysis and control of constrained particulate processes of the form of Eq.5.1:

1 Initially, the method of weighted residuals is used to derive a nonlinear ODE system that accurately reproduces the solutions and dynamics of the system of Eq.5.1.

2 Next, the low-order ODE approximation of the system of Eq.5.1 is used as the basis for analyzing the fundamental limitations imposed by input constraints on our ability to modify the dynamics of the particulate process. This is done by identifying the set of admissible set-points that can be attained in the presence of the given input constraints.

3 Then, given the set of feasible control objectives, the low-order ODE approximation of the system of Eq.5.1 is used as the basis for the direct synthesis, via Lyapunov techniques, of practically implementable bounded nonlinear output feedback controllers that cope effectively with the problem of input constraints by: a) enforcing stability and set-point tracking in the constrained closed-loop ODE system , and b) providing an explicit characterization of the set of operating conditions starting from where the desired cons closed-loop stability and performance are guaranteed in the presence of constraints.

4 Finally, the resulting closed-loop system (particulate process model of Eq.5.1 and controller) is analyzed to derive conditions that guarantee that the desired stability and set-point tracking properties are enforced in the infinite-dimensional closed-loop system.

4. Model Reduction

We initially use the method of weighted residuals to derive a set of nonlinear ODEs that accurately reproduce the solutions and the dominant dynamics of the distributed parameter system of Eq.5.1. The central idea of the method of weighted residuals is to approximate the exact solution of $n(r,t)$ by an infinite series of orthogonal basis functions defined on the interval $[0, r_{max})$ with time-varying coefficients, substitute the series expansion into Eq.5.1, and then take the inner product with respect to a complete set of weighted functions, to compute a set of ODEs which describes the rate of change of the time-varying coefficients of the series expansion of the solution. Specifically, we expand the solution of $n(r,t)$ in an infinite series in terms of an orthogonal and complete set of basis functions, $\phi_k(r)$, where $r \in [0, r_{max})$, $k = 1, \ldots, \infty$, as follows:

$$n(r,t) \quad = \quad \sum_{k=1}^{\infty} a_k(t)\phi_k(r) \tag{5.7}$$

where $a_k(t)$ are time-varying coefficients. Substituting the above expansion into the particulate process model of Eq.5.1, we obtain:

$$\begin{aligned}
\sum_{k=1}^{\infty} \phi_k(r) \frac{\partial a_k(t)}{\partial t} &= -\sum_{k=1}^{\infty} a_k(t) \frac{\partial (G(x,r)\phi_k(r))}{\partial r} + w(\sum_{k=1}^{\infty} a_k(t)\phi_k(r), x, r) \\
\dot{x} &= f(x) + g(x)sat(u(t)) + A\int_0^{r_{max}} a_2(\sum_{k=1}^{\infty} a_k(t)\phi_k(r), r, x)dr)
\end{aligned} \tag{5.8}$$

Multiplying the population balance with the weighting functions, $\psi_\nu(r)$, and integrating over the entire particle size spectrum (i.e., taking inner product in $L_2[0, r_{max})$ with the weighting functions), the following infinite set of nonlinear ODEs is obtained:

$$\int_0^{r_{max}} \psi_\nu(r) \sum_{k=1}^{\infty} \phi_k(r) \frac{\partial a_k(t)}{\partial t} dr = -\sum_{k=1}^{\infty} a_k(t) \int_0^{r_{max}} \psi_\nu(r) \frac{\partial (G(x,r)\phi_k(r))}{\partial r} dr$$
$$+ \int_0^{r_{max}} \psi_\nu(r) w(\sum_{k=1}^{\infty} a_k(t)\phi_k(r), x, r) dr, \quad \nu = 1, \dots, \infty$$

$$\dot{x} = f(x) + g(x) sat(u(t)) + A \int_0^{r_{max}} a_2(\sum_{k=1}^{\infty} a_k(t)\phi_k(r), r, x) dr) \tag{5.9}$$

Eq.5.9 is an infinite set of ODEs which describe the rate of change of the time varying coefficients, $a_k(t)$, where $k = 1, \dots, \infty$, of the series expansion of the solution. An accurate approximation of Eq.5.9 is obtained by truncating the series expansion of $n(r, t)$ up to order N and keeping the first N equations (i.e., $\nu = 1, \dots, N$). The infinite-dimensional system of Eq.5.9 reduces to the following finite set of ODEs:

$$\int_0^{r_{max}} \psi_\nu(r) \sum_{k=1}^{N} \phi_k(r) \frac{\partial a_{kN}(t)}{\partial t} dr$$
$$= -\sum_{k=1}^{N} a_{kN}(t) \int_0^{r_{max}} \psi_\nu(r) \frac{\partial (G(x_N,r)\phi_k(r))}{\partial r} dr$$
$$+ \int_0^{r_{max}} \psi_\nu(r) w(\sum_{k=1}^{N} a_{kN}(t)\phi_k(r), x_N, r) dr, \quad \nu = 1, \dots, N$$

$$\dot{x}_N = f(x_N) + g(x_N) sat(u(t)) + A \int_0^{r_{max}} a_2(\sum_{k=1}^{N} a_{kN}(t)\phi_k(r), r, x_N) dr) \tag{5.10}$$

where x_N and a_{kN} are the approximations of x and a_k obtained by an N-th order truncation. Introducing the vector notation $a_N = [a_{1N} \; \cdots \; a_{NN}]$, and after some rearrangements, Eq.5.10 can be represented in the following general form:

$$\begin{aligned} \dot{a}_N &= f^*(a_N, x_N) \\ \dot{x}_N &= f(x_N) + g(x_N) sat(u(t)) + A \int_0^{r_{max}} a_2(\sum_{k=1}^{N} a_{kN}(t)\phi_k(r), r, x_N) dr) \end{aligned} \tag{5.11}$$

where the explicit expression of $f^*(a_N, x_N)$ is omitted for brevity. Setting $\tilde{x} = [a_N^T \; x_N^T]^T$, we obtain the following multi-input multi-output finite-dimensional ODE system:

$$\begin{aligned} \dot{\tilde{x}} &= \tilde{f}(\tilde{x}) + \sum_{i=1}^{m} \tilde{g}_i(\tilde{x}) sat(u_i) \\ y_{si} &= \tilde{h}_i(\tilde{x}), \quad i = 1, \ldots, m \end{aligned} \tag{5.12}$$

where $\tilde{f}(\tilde{x}), \tilde{g}_i(\tilde{x}), \tilde{w}(\tilde{x})$ are nonlinear vector functions whose explicit form is omitted brevity. Using results from singular perturbation theory, the asymptotic validity of the ODE approximation was established in Proposition 2.1.

5. Computation of Admissible Set-points

Having obtained a low-order approximate ODE system that accurately reproduces the solution and dominant dynamics of the particulate process, we are now in a position to analyze the limitations imposed by constraints on our ability to modify the dynamics of particulate processes, on the basis of the constrained ODE system of Eq.5.12. Particulate processes typically operate at constant set-points corresponding to equilibrium points of the closed-loop system (e.g., a desired total number of particles, mean particle size, temperature). However, it may not be feasible to steer the closed-loop system to the desired operating point, in the presence of constraints, irrespectively of the choice of the control strategy. Thus, even before designing a control policy, it becomes important to investigate the feasibility of controlling the particulate process at a desired set-point in the presence of constraints. Addressing this problem, in its full generality, entails two main tasks. The first task is that of identifying the set of admissible set-points that the particulate process can be steered to in the presence of constraints. The second task is that of characterizing the set of admissible initial conditions, starting from where, a given admissible set-point can be reached in the presence of constraints. In this section, we will focus on the first task and defer discussion of the second task to the next section.

To accomplish our goal of identifying the set of admissible set-points permitted by input constraints, it is useful to view the constrained ODE system of Eq.5.12 as a dynamical system where the control input u_i is viewed as a parameter that takes values in the set of admissible control inputs, i.e. the interval $U_i = [u_{i,min}, u_{i,max}]$. In other words, we consider the following unforced system

$$\dot{\tilde{x}} = \tilde{f}(\tilde{x}) + \sum_{i=1}^{m} \tilde{g}_i(\tilde{x}) u_i^0, \quad i = 1, \ldots, m \tag{5.13}$$

obtained from the system of Eq.5.12 for a constant value of u_i denoted by $u_i^0 \in U_i$. An immediate consequence of this view is the realization that the presence of constraints on the values that the parameter u_i^0 in the system of Eq.5.13 can take leads to natural limitations on the steady-states or equilibrium states and, consequently the set-points, that the ODE system can attain. Exploiting this fact, the problem of characterizing the limitations imposed by input constraints is equivalent to that of explicitly characterizing the dependence of the equilibrium states of the constrained ODE system of Eq.5.13 on the admissible values of u_i^0 dictated by the given input constraints. For a given value of the control input $u_i^0 \in U_i$, one can obtain the admissible equilibrium states of the system of Eq.5.12 by solving the following set of algebraic equations

$$0 = \tilde{f}(\tilde{x}_s) + \sum_{i=1}^{m} \tilde{g}_i(\tilde{x}_s)u_i^0, \quad i = 1, \ldots, m \tag{5.14}$$

where $\tilde{x}_s$ represents the admissible steady-state for the above ODE system. As the value of u_i^0 is varied smoothly over the set of admissible control inputs U_i, we obtain the set of all admissible equilibrium points of the constrained ODE system of Eq.5.12 that can be attained in the presence of constraints. Finally, the admissible set-points can be computed directly from the following relation:

$$v_i = \tilde{h}_i(\tilde{x}_s), \quad i = 1, \ldots, m \tag{5.15}$$

Remark 5.1: The above analysis provides a systematic and practical method for computing the approximate admissible set-points of the constrained particulate process of Eq.5.1 on the basis of the ODE system of Eq.5.12. The only task involved in this method is the solution of a small (due to the low-dimensional nature) set of algebraic equations of the form of Eq.5.14 as u_i^0 takes values in the set of admissible control inputs. Note that this analysis is independent of the specific control strategy that one may wish to implement. Note also that this analysis does not provide any information regarding where, in state-space, a particular admissible set-point can be reached from. In other words, given an admissible equilibrium point $\tilde{x}_s$ obtained by solving Eq.5.14, the above analysis does not say if this point will actually be achieved under a given controller and starting from a given initial condition. The task of providing such information will be addressed in the controller design step of our methodology. What the above analysis allows us to conclude, however, is whether the given input constraints place any fundamental limitations on our ability to reach a particular set-point, and can therefore be used by process operators to identify a priori (before controller design) admissible control objectives.

Remark 5.2: Recent research on the dynamical analysis of nonlinear control systems with input constraints (see, for example, Colonius and Kliemann, 1993) has provided a system-theoretic characterization of the regions of controllability under constraints, the so-called control sets, where *any* two points in state-space can be reached from each other with the available control action. As the range of manipulated input is varied smoothly over a sufficiently small finite range, these control sets emerge and evolve around the equilibrium points of the nominal system of Eq.5.13. Although control sets provide a more detailed characterization of the limitations imposed by input constraints on the system dynamics, such characterization remains largely theoretical and not very useful from a computational standpoint. The construction of control sets is actually a very cumbersome task even for low-dimensional systems (see, for example, Kapoor and Daoutidis, 1999.) On the other hand, computation of the set of admissible set-points, contained within a control set, is a relatively straightforward task that is computationally feasible and provides at the same time a valuable piece of information. It is noteworthy here that the concept of using steady-state system analysis to reach certain conclusions about the control of a dynamic system has been discussed in the context of integration of process design and control (e.g., Stephanopoulos, 1983; Fisher et al., 1985).

Let us now apply our feasibility analysis to the continuous crystallizer example considered earlier in the preliminaries section. It was shown in Chapter 2 that upon application of the method of moments to the continuous crystallizer process of Eq.5.4 and neglecting moments of order four and higher, one can derive the following dimensionless ODE system that accurately reproduces the dynamics of the distributed process model of Eq.5.4:

$$\begin{aligned}
\dot{\tilde{x}}_0 &= -\tilde{x}_0 + (1 - \tilde{x}_3) Da e^{\frac{-F}{\tilde{y}^2}} \\
\dot{\tilde{x}}_1 &= -\tilde{x}_1 + \tilde{y}\tilde{x}_0 \\
\dot{\tilde{x}}_2 &= -\tilde{x}_2 + \tilde{y}\tilde{x}_1 \\
\dot{\tilde{x}}_3 &= -\tilde{x}_3 + \tilde{y}\tilde{x}_2 \\
\dot{\tilde{y}} &= \frac{1 - \tilde{y} - (\alpha - \tilde{y})\tilde{y}\tilde{x}_2}{1 - \tilde{x}_3} + \frac{u}{1 - \tilde{x}_3}
\end{aligned} \tag{5.16}$$

where

$$t = \frac{\bar{t}}{\tau}, \; \tilde{x}_0 = 8\pi\sigma^3\mu_0, \; \tilde{x}_1 = 8\pi\sigma^2\mu_1, \; \tilde{x}_2 = 4\pi\sigma\mu_2, \; \tilde{x}_3 = \frac{4}{3}\pi\mu_3, \; \ldots,$$
$$\sigma = k_1\tau(c_0 - c_s), \; Da = 8\pi\sigma^3 k_2\tau, \; F = \frac{k_3 c_s^2}{(c_0 - c_s)^2}, \; \alpha = \frac{(\rho - c_s)}{(c_0 - c_s)},$$
$$\tilde{y} = \frac{(c - c_s)}{(c_0 - c_s)}, \; u = \frac{(c_0 - c_{0s})}{(c_0 - c_s)} \tag{5.17}$$

μ_i is the $i-$th moment, $\tilde{x}_0$ is a dimensionless crystal concentration and u is a dimensionless solute feed concentration (for definitions of the remaining variables and values of the model parameters, the reader is referred to Chapter 2). To compute the admissible set-points for the above system, we take the set of admissible control inputs to be $U = [0, 6]$. Setting the left hand side of Eq.5.16 equal to zero, we obtain a set of algebraic equations that can be solved for each value of $u \in U$. Figure 5.3 illustrates the

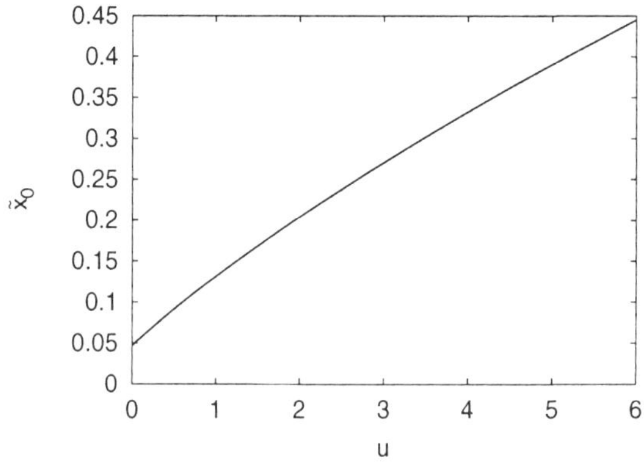

Figure 5.3. Admissible set-points for crystal concentration for $u \in [0, 6]$.

admissible set-points for the crystal concentration which can be achieved with the available control inputs. From this figure, the reason for the PI controller's failure to stabilize the crystallizer in section 2.2 and achieve the requested 0.4 change in the set point is evident. Allowing the control inputs to vary only in the set $[0, 2]$ renders the requested set-point $\tilde{x}_0 = 0.4$ infeasible.

6. Bounded Nonlinear Control of Particulate Processes

Having obtained a low-order ODE system that captures the dominant dynamics of the particulate process and identified the set of feasible control objectives that can be achieved in the presence of constraints, we are now motivated to proceed with the third step of our proposed methodology. In this section, we use the constrained low-order ODE system of Eq.5.12 as the basis for developing an effective control strategy that handles explicitly the problem of constraints. The key components of this strategy involve: a) the synthesis of a practically implementable bounded nonlinear output feedback controller that enforces stability and set-point tracking in the closed-loop system in the presence of active input constraints and b) the explicit characterization of the state-space region of guaranteed closed-loop stability associated with the designed controller. The output feedback controller is constructed through a standard combination of a bounded state feedback controller with a state observer. The state feedback controller is synthesized via Lyapunov-based control methods and the state observer is an extended Luenberger-type observer. Before we proceed with the controller design, we begin in the next subsection with some preliminaries that will be used to state the controller synthesis result.

6.1 Preliminaries

Referring to the system of Eq.5.12, we define the relative order of the output y_{si} with respect to the vector of manipulated inputs u as the smallest integer r_i for which

$$\left[L_{\tilde{g}_1} L_{\tilde{f}}^{r_i-1} \tilde{h}_i(\tilde{x}) \ \cdots \ L_{\tilde{g}_m} L_{\tilde{f}}^{r_i-1} \tilde{h}_i(\tilde{x})\right] \not\equiv [0 \ \cdots \ 0] \tag{5.18}$$

or $r_i = \infty$ if such an integer does not exist. We also define the characteristic matrix

$$C(\tilde{x}) = \begin{bmatrix} L_{\tilde{g}_1} L_{\tilde{f}}^{r_1-1} \tilde{h}_1(\tilde{x}) & \cdots & L_{\tilde{g}_m} L_{\tilde{f}}^{r_1-1} \tilde{h}_1(\tilde{x}) \\ L_{\tilde{g}_1} L_{\tilde{f}}^{r_2-1} \tilde{h}_2(\tilde{x}) & \cdots & L_{\tilde{g}_m} L_{\tilde{f}}^{r_2-1} \tilde{h}_2(\tilde{x}) \\ \vdots & \cdots & \vdots \\ L_{\tilde{g}_1} L_{\tilde{f}}^{r_m-1} \tilde{h}_m(\tilde{x}) & \cdots & L_{\tilde{g}_m} L_{\tilde{f}}^{r_m-1} \tilde{h}_m(\tilde{x}) \end{bmatrix} \tag{5.19}$$

To proceed with the controller synthesis and under the assumption that the relative degree is well-defined, we transform the system of Eq.5.12

into the following partially linear form

$$
\begin{array}{rcl}
\dot{\zeta}_1^{(i)} & = & \zeta_2^{(i)} \\
& \vdots & \\
\dot{\zeta}_{r_i-1}^{(i)} & = & \zeta_{r_i}^{(i)} \\
\dot{\zeta}_{r_i}^{(i)} & = & L_{\tilde{f}}^{r_i}\tilde{h}_i(\tilde{x}) + \sum_{k=1}^{m} L_{\tilde{g}_k} L_{\tilde{f}}^{r_i-1}\tilde{h}_i(\tilde{x})u_k \\
\dot{\eta}_1 & = & \Psi_1(\zeta,\eta) \\
& \vdots & \\
\dot{\eta}_{(n+N)-\sum_i r_i} & = & \Psi_{(n+N)-\sum_i r_i}(\zeta,\eta) \\
y_{si} & = & \zeta_1^{(i)}, \quad i=1,\ldots,m
\end{array}
\tag{5.20}
$$

where $\tilde{x} = T^{-1}(\zeta,\eta)$, $\zeta = [\zeta^{(1)^T} \ \cdots \ \zeta^{(m)^T}]^T$. This transformation is standard in all general nonlinear process control methods whose objective is to force the process output to follow the reference input. Defining the tracking error $e_k^{(i)} = \zeta_k^{(i)} - v_i^{(k-1)}$ and introducing the vector notation $e^{(i)} = [e_1^{(i)} \ e_2^{(i)} \ \cdots \ e_{r_i}^{(i)}]^T$, $e = [e^{(1)^T} \ e^{(2)^T} \ \cdots \ e^{(m)^T}]^T$, where $i = 1, \cdots, m$, $k = 1, \cdots, r_i$, the ζ-subsystem of Eq.5.20 can be further transformed into the following more compact form

$$\dot{e} \ = \ Ae + B\left[l_1(e,\eta,\bar{v}) + C(\tilde{x})u\right] \tag{5.21}$$

where A, B, are constant matrices of dimensions $(\sum_{i=1}^m r_i)\times(\sum_{i=1}^m r_i)$ and $(\sum_{i=1}^m r_i)\times m$, respectively, $l_1(e,\eta,\bar{v})$ is a $(\sum_{i=1}^m r_i)\times 1$ continuous nonlinear vector function, and $\bar{v}$ is a vector of the form $\bar{v} = \mathcal{V}(v_i, v_i^{(1)}, \ldots, v_i^{(r_i)})$ where $\mathcal{V}(v_i, v_i^{(1)}, \ldots, v_i^{(r_i)})$ is a smooth vector function, $v_i^{(k)}$ is the kth time derivative of the external reference input v_i (which is assumed to be a smooth function of time.) The specific forms of these functions are omitted for brevity. Finally, we define the function $\bar{f}(e,\eta,\bar{v}) = Ae + Bl_1(\tilde{x})$, and denote by $\bar{g}_i(e,\eta,\bar{v})$ the $i-$th column of the matrix function $BC(\tilde{x})$, $\quad i = 1, \cdots m$.

6.2 Controller Synthesis

Towards the end goal of synthesizing the necessary bounded nonlinear output feedback controller, we use the nonlinear system of Eq.5.20 first to synthesize, via Lyapunov-based control methods, a bounded nonlinear state feedback controller of the general form

$$u \ = \ p(\tilde{x},\bar{v}) \tag{5.22}$$

where $p(\tilde{x}, \bar{v})$ is a bounded vector function, i.e. $|u| \leq u_{max}$ where $|\cdot|$ is the Euclidean norm, that: a) enforces exponential stability and reference input tracking in the closed-loop system in the presence of active input constraints and b) provides an explicit characterization of the region in state-space where the aforementioned closed-loop properties are guaranteed.

To construct the desired stabilizing state feedback controller for the system of Eq.5.12, we use Lyapunov-based control methods. The basic idea behind any Lyapunov-based controller design is the selection of an appropriate Lyapunov function whose time-derivative can be rendered negative definite, via feedback, along the trajectories of the closed-loop system. A natural choice for our system, suggested by the partially linear form of Eq.5.21, is a quadratic Lyapunov function $V = e^T Pe$ where P is a positive definite matrix chosen to satisfy the following Riccati matrix inequality

$$A^T P + PA - PBB^T P \;<\; 0 \tag{5.23}$$

which consequently guarantees the negative definiteness of $\dot{V}$. Using this Lyapunov function, we design a bounded nonlinear state feedback controller of the form

$$u \;=\; -\frac{1}{2} R^{-1}(\tilde{x})\,(L_{\bar{g}}V)^T \tag{5.24}$$

where $R^{-1}(\tilde{x})$ is a strictly positive nonlinear scalar function whose specific expression is given in theorem 1 below and $L_{\bar{g}}V$ is a row vector given by $L_{\bar{g}}V = [L_{\bar{g}_1}V \cdots L_{\bar{g}_m}V]$. Specifically, the design of the bounded nonlinear state feedback controller will be done by appropriately modifying the bounded controller design proposed by Lin and Sontag, 1991, to ensure that the constrained finite-dimensional closed-loop system is locally exponentially stable (see also discussion in remark 5.6 below). As we will discuss shortly, the bounded nature of the state feedback controller will assist in addressing the problem of constraints by providing an explicit characterization of the set of admissible initial states, starting from where, the desired closed-loop properties are guaranteed.

Under the hypothesis that the system of Eq.5.12 is locally observable (i.e., its linearization around the desired operating steady-state is observable), the practical implementation of a nonlinear state feedback controller of the form of Eq.5.24 will be achieved by employing the following nonlinear state observer:

$$\frac{d\omega}{dt} \;=\; \tilde{f}(\omega) + \tilde{g}(\omega)sat(u) + L(y - \tilde{h}(\omega)) \tag{5.25}$$

where ω denotes the observer state vector (the dimension of the vector ω is equal to the dimension of $\tilde{x}$ in the system of Eq.5.12), $y =$

$[y_1\ y_2\ \cdots y_l]^T$ is the measured output vector and L is a matrix chosen so that the eigenvalues of the matrix $C_L = \frac{\partial \tilde{f}}{\partial \omega}_{(\omega=\omega_s)} - L\frac{\partial \tilde{h}}{\partial \omega}_{(\omega=\omega_s)}$, where ω_s is the operating steady-state, lie in the open left-half of the complex plane. The state observer of Eq.5.25 consists of a replica of the system of Eq.5.12 plus a linear gain multiplying the discrepancy between the actual and the estimated value of the output, and therefore, it is an extended Luenberger-type observer. Finally, the state feedback control law of Eq.5.24 and the state observer of Eq.5.25 can be combined to yield the desired bounded nonlinear output feedback control law.

We are now in a position to state the main result of this subsection. Theorem 5.1 below provides the explicit synthesis formula for the desired bounded output feedback control law and states precise conditions that guarantee closed-loop stability and asymptotic output tracking in the presence of input constraints in the closed-loop ODE system (the proof can be found in appendix 9.C.)

Theorem 5.1: *Consider the system of Eq.5.12 and assume that:* 1) *it is locally observable in the sense that there exists a matrix L such that* $C_L = \frac{1}{\mu}\bar{A}$ *where* μ *is a small positive parameter and* $\bar{A}$ *is a Hurwitz matrix,* 2) *its characteristic matrix,* $C(\tilde{x})$*, is nonsingular* $\forall\ \tilde{x} \in D \subset \mathbb{R}^{n+N}$, *and* 3) *its inverse dynamics are input-to-state stable and exponentially stable when* $\zeta = 0$. *Consider first the system of Eq.5.12 under the state feedback controller of Eq.5.24 where*

$$\frac{1}{2}R^{-1}(\tilde{x}) = \frac{L_{\bar{f}}^{*}V + \sqrt{(L_{\bar{f}}^{*}V)^2 + (u_{max}^2(L_{\bar{g}}V)(L_{\bar{g}}V)^T)^2}}{(L_{\bar{g}}V)(L_{\bar{g}}V)^T\left[1+\sqrt{1+u_{max}^2(L_{\bar{g}}V)(L_{\bar{g}}V)^T}\right]} \tag{5.26}$$

and $L_{\bar{f}}^{*}V = L_{\bar{f}}V + \rho|e|^2$, $\rho > 0$. *Let* δ_x *be a positive real number such that the compact set* $|\tilde{x}| \leq \delta_x$ *is the largest invariant set embedded within the unbounded region described by the following inequality*

$$L_{\bar{f}}^{*}V \leq u_{max}|(L_{\bar{g}}V)^T| \tag{5.27}$$

Then, for any initial condition that satisfies $|\tilde{x}(0)| \leq \delta_x$, *the closed-loop system under state feedback control is asymptotically stable in the sense that there exists a function* β *of class KL such that* $|x(t)| \leq \beta(|x(0)|, t)$, $\forall t \geq 0$. *Now consider the system of Eq.5.12 under the output feedback controller*

$$\begin{aligned}\frac{d\omega}{dt} &= \tilde{f}(\omega) - \frac{1}{2}\tilde{g}(\omega)R^{-1}(\omega)\left(L_{\bar{g}}V(\omega)\right)^T + L(y - \tilde{h}(\omega)) \\ u &= -\frac{1}{2}R^{-1}(\omega)\left(L_{\bar{g}}V(\omega)\right)^T\end{aligned} \tag{5.28}$$

Then given any pair of positive real numbers (d, δ_b) such that $\beta(\delta_b, 0) + d \leq \delta_x$, there exists $\mu^* > 0$ such that if $\mu \in (0, \mu^*]$, $|\tilde{x}(0)| \leq \delta_b$, $|\omega(0)| \leq \delta_b$, the following holds in the presence of input constraints: 1) the closed-loop system is asymptotically (and locally exponentially) stable, 2) the output of the closed-loop ODE system satisfies a relation of the form $\lim_{t\to\infty}|y_{si} - v_i| = 0$, where v_i is the set-point for the ith controlled output.

Remark 5.3: Theorem 5.1 provides an explicit characterization of a set of admissible initial conditions starting from where the constrained closed-loop ODE system is guaranteed to be stable with the available control action (region of guaranteed closed-loop stability). This characterization can be obtained from the inequality of Eq.5.27. This inequality describes the largest region in state space where $\dot{V}$ is negative definite under the bounded state feedback controller of Eq.5.24-5.26 and where any closed-loop trajectory evolving is guaranteed to converge to the desired equilibrium point with the available control action. However, since this region is, in general, not an invariant one, it is necessary to guarantee that trajectories starting within the region do not leave, in order to guarantee closed-loop stability. This is done by confining the initial conditions within the largest invariant set embedded within the region. The size of this set is fixed by δ_x and represents the state feedback estimate of the region of closed-loop stability (see El-Farra and Christofides, 2001a and chapter 4 in Khalil, 1996 for how to compute this estimate). According to theorem 5.1, this estimate remains practically preserved under output feedback. Specifically, starting from any compact subset (whose size is fixed by δ_b) of the state feedback region, there always exists $\mu > 0$ such that the desired closed-loop properties are guaranteed under the dynamic *output* feedback controller of Eq.5.28. Note that the size of the output feedback region (δ_b) can be made close to that of the state feedback region (δ_x) by selecting d to be sufficiently small which, in turn, can be done by choosing μ to be sufficiently small. Therefore, although combination of the bounded state feedback controller with an observer results in some loss (represented by d) in the size of the region of guaranteed closed-loop stability, this loss can be made arbitrarily small by selecting μ to be sufficiently small.

Remark 5.4: The inequality of Eq.5.27 captures, in an intuitive way, the dependence of the size of the region of guaranteed closed-loop stability on the magnitude of input constraints. For example, this inequality predicts that the tighter the input constraints are made (i.e., smaller u_{max}), the smaller the resulting closed-loop stability region. This is consistent with one's intuition, since under tight constraints, only few initial conditions are admissible and can be used to stabilize the closed-loop system. Finally note that, according to the inequality of Eq.5.27, the largest region of closed-loop stability under the control law of Eq.5.24 is obtained, as expected, in the absence of constraints (i.e., as $u_{max} \to \infty$).

Remark 5.5: Although the inequality of Eq.5.27 can be used to compute the region of guaranteed closed-loop stability, the actual construction of this region is not necessary to conclude whether the desired closed-loop properties can be guaranteed for a given initial condition. Instead, one can directly substitute the desired initial condition into Eq.5.27 and check before implementing the controller whether the inequality is satisfied. If so, then both stability and set-point tracking are guaranteed in the presence of input constraints. This aspect of the controller design has important practical implications because it provides process operators with a systematic and easily implementable guide to identify a priori (that is before implementing the controller) feasible initial operating conditions form which to operate and steer the process safely and reliably under input constraints while at the same time satisfying the requested control objectives.

Remark 5.6: Note that the static state feedback component of the proposed output feedback controller of Eq.5.28 (with $\omega = \tilde{x}$) uses directly the available information on input constraints to achieve the requested closed-loop stability and performance properties (note the explicit dependence of the expression in Eq.5.26 on u_{max}.) Note also that, whenever Eq.5.27 holds, the control action is bounded by u_{max} and satisfies the constraints. Both the explicit incorporation of constraints in the controller design and the explicit characterization of the region of guaranteed closed-loop stability follow directly from the inherent boundedness property of the state feedback component in Eq.5.28. It is important to point out here that the design of the state feedback controller of Eq.5.24-5.26 involves a modification of the controller design proposed in Lin and Sontag, 1991 by introducing the term $-\rho|e|^2$ which is necessary to ensure that the constrained finite-dimensional closed-loop system is locally exponentially stable.

Remark 5.7: The control law of Eq.5.28 has two desirable properties not present in other conventional or nonlinear controller designs based on the concept of input/output linearization (see Isidori, 1995 for details.) The first property is the fact that the controller of Eq.5.28 recognizes the beneficial effect of the term L_fV when $L_fV < 0$ and prevents its unnecessary cancellation. In this case, the term L_fV is a beneficial (stabilizing) nonlinearity whose cancellation may generate unnecessarily large control action, and in some cases generate positive feedback and destabilize the process. Under such circumstances, the controller design of Eq.5.28 guards against the wasteful cancellation of such nonlinearities and avoids the expenditure of unnecessarily large control effort. This property is particularly important in the presence of constraints that limit the available control action. The ability of the controller to use reasonably small control action to enforce the desired control objectives makes the controller design proposed in theorem 1 better equipped to handle the problem of constraints. The second property that the controller of Eq.5.28 possesses is the fact that it dominates the term L_fV, instead of cancelling it, when $L_fV > 0$. This situation arises, for example, when the process is open-loop unstable. In this case, the term L_fV is a destabilizing one that must be eliminated. The controller of Eq.5.28, however, eliminates the term by domination rather than by cancellation. This property guards against the non-robustness of cancellation designs which increases the risk of instability due to the presence of other uncertainties not taken into account in the controller design. This means that input perturbations (or equivalently, an error in implementing the control law) will be tolerated in the sense that the trajectories will remain bounded.

Remark 5.8: The bounded nonlinear controller of Eq.5.24 possesses certain optimality properties characteristic of its ability to use small control action to accomplish the desired closed-loop objectives. In fact, one can rigorously prove, through the inverse optimal control approach (see, for example, Freeman and Kokotovic, 1996; Sepulchre et al., 1997; El-Farra and Christofides, 2001a), that, within a well-defined region of the closed-loop stability region, this controller is optimal with respect to an infinite-time meaningful quadratic cost functional of the form

$$J = \int_0^{\infty} (e^T Q(\tilde{x}) e + u^T R(\tilde{x}) u) dt \qquad (5.29)$$

which imposes penalty on both the tracking error and control action. In the above performance index, $Q(\tilde{x})$ is a positive definite matrix that can be found directly from the steady-state Hamilton-Jacobi-Bellman (HJB) equation (which is the optimality condition for the stabilization problem) associated with the system of Eq.5.12 and cost functional of Eq.5.29

$$0 \equiv e^T Q(\tilde{x})e + L_{\tilde{f}}V - \frac{1}{4}(L_{\tilde{g}}V)R^{-1}(\tilde{x})\,(L_{\tilde{g}}V)^T \tag{5.30}$$

The inverse optimal approach provides a rigorous framework for associating meaningful optimality (i.e. meaningful performance indices) with certain stabilizing controllers (such as those of Eq.5.24) and therefore helps explain the basis for their optimality properties. The key idea of this approach is to first design a stabilizing controller and then show that it minimizes a meaningful performance index of the form of Eq.5.29 by establishing that the resulting weights Q, R are positive definite, which, in turn, renders the penalties imposed in the cost functional physically sensible. This approach has been used in the literature for the design of nonlinear optimal controllers without recourse to the unwieldy task of solving the HJB equation. For additional details on controller design using this approach as well as some of the history and motivation behind it, the reader is referred to Sepulchre et al., 1997; El-Farra and Christofides, 2001a. Finally, one can easily show that the minimum cost achieved by the state feedback controller is $V(e(0))$.

Remark 5.9: The controller-observer combination of Eq.5.28 practically preserves the optimality properties of the state feedback controller explained in the previous remark. The output feedback controller design is near-optimal in the sense that the cost incurred by implementing this controller on the system of Eq.5.12 tends to the optimal (minimal) cost achieved by implementing the bounded optimal state feedback controller (i.e., u of Eq.5.28 with $\omega = \tilde{x}$) when μ is selected to be sufficiently small. Using a standard singular perturbation argument, one can show that cost associated with the output feedback controller is $O(\mu)$ close to the optimal cost associated with the state feedback controller (i.e. $J_{min} = V(e(0)) + O(\mu)$). The basic reason for near-optimality is the fact that by choosing μ to be sufficiently small, the observer states can be made to converge quickly to the process states. This fact can be exploited to make the performance of the output feedback controller arbitrarily close to that of the optimal state feedback controller.

Remark 5.10: The requirement that the observer states start within the region of guaranteed closed-loop stability ($|\omega(0)| \leq \delta_b$) is motivated by the fact that the process states themselves should start inside this region in order to guarantee stability of the constrained closed-loop system; and therefore, one must initiate the observer within the same region to guarantee convergence of the observer states to the process states. Note, however, that no restriction is placed on where, inside the stability region, the observer states can start (that is we allow for initialization errors) since they can be made to converge sufficiently fast to the actual states by selecting μ to be sufficiently small.

Remark 5.11: Regarding the practical application of theorem 5.1, one has to initially use the method of weighted residuals to derive an ODE system of the form of Eq.5.12, and then verify assumptions 1, 2 and 3 of the theorem on the basis of this system. Then, given the available input constraints u_{max} and the desired initial condition, one should check if the inequality of Eq.5.27 is satisfied and if the initial condition lies within the region of guaranteed closed-loop stability. If this is case, then the desired closed -loop properties are guaranteed and the synthesis formula of Eq.5.28 can be directly used to derive the explicit form of the controller and implement it.

6.3 Application to the Crystallizer Moment Model

The objective of this subsection is to illustrate the application of the results of Theorem 5.1 to the fifth-order moment model of Eq.5.16 which describes the dominant dynamics of the continuous crystallizer of Eq.5.4. The control problem here is the same as that considered in the motivating example which involves regulating the crystal concentration by manipulating the solute feed concentration in the presence of constraints on the manipulated input. Utilizing the dimensionless variables of Eq.5.17, the system of Eq.5.16 can be recast in the form of Eq.5.12 with $\tilde{x} = [\tilde{x}_0 \; \tilde{x}_1 \; \tilde{x}_2 \; \tilde{x}_3 \; \tilde{y}]^T$ and

$$\tilde{f}(\tilde{x}) = \begin{bmatrix} -\tilde{x}_0 + (1-\tilde{x}_3)Dae^{\frac{-F}{\tilde{y}^2}} \\ -\tilde{x}_1 + \tilde{y}\tilde{x}_0 \\ -\tilde{x}_2 + \tilde{y}\tilde{x}_1 \\ -\tilde{x}_3 + \tilde{y}\tilde{x}_2 \\ \dfrac{1-\tilde{y}-(\alpha-\tilde{y})\tilde{y}\tilde{x}_2}{1-\tilde{x}_3} \end{bmatrix}, \quad \tilde{g}(\tilde{x}) = \begin{bmatrix} 0 \\ 0 \\ 0 \\ 0 \\ \dfrac{1}{1-\tilde{x}_3} \end{bmatrix}$$

On the basis of this system, one can easily verify that assumptions 1, 2 and 3 of Theorem 5.1 are satisfied. A direct application of the synthesis formula of Eq.5.28 then yields the following bounded nonlinear output feedback controller

$$
\begin{aligned}
\dot{\omega}_0 &= -\omega_0 + (1-\omega_3)Dae^{\frac{-F}{\omega_4^2}} + L_0[\tilde{x}_0 - \omega_0] \\
\dot{\omega}_1 &= -\omega_1 + \omega_4\omega_0 + L_1[\tilde{x}_0 - \omega_0] \\
\dot{\omega}_2 &= -\omega_2 + \omega_4\omega_1 + L_2[\tilde{x}_0 - \omega_0] \\
\dot{\omega}_3 &= -\omega_3 + \omega_4\omega_2 + L_3[\tilde{x}_0 - \omega_0] \\
\dot{\omega}_4 &= \frac{1-\omega_4-(\alpha-\omega_4)\omega_4\omega_2}{1-\omega_3} - \frac{1}{2}\frac{R^{-1}(\omega)L_{\bar{g}}V(\omega)}{1-\omega_3} + L_4[\tilde{x}_0 - \omega_0] \\
u &= -\frac{1}{2}R^{-1}(\omega)L_{\bar{g}}V(\omega)
\end{aligned}
\tag{5.31}
$$

where

$$
V = e^T P e, \quad P = \begin{bmatrix} 1 & c \\ c & 1 \end{bmatrix}, \quad c \in (0,1) \tag{5.32}
$$

Guided by the information provided in Figure 5.3, we constrain the admissible values of the manipulated input to vary in the range $0 \leq u \leq 6$ to achieve the requested set-point of $\tilde{x}_0 = 0.4$. Two closed-loop simulation runs were performed to evaluate the set-point tracking capability of the bounded nonlinear output feedback controller of Eq.5.31 in the presence of constraints. The initial conditions for the states of the moment model, the states of the observer, and the values of the nonlinear controller parameters are all given in Table 5.1 for both runs.

	First run	Second run
$\tilde{x}(0)$	$[0.059\ 0.035\ 0.022\ 0.014\ 0.60]^T$	$[0.44\ 0.61\ 0.85\ 1.14\ 0.60]^T$
$\omega(0)$	$[0.047\ 0.028\ 0.017\ 0.010\ 0.60]^T$	$[0.42\ 0.59\ 0.80\ 1.07\ 0.60]^T$
L	$[1.0\ 0.0\ 0.0\ 0.0\ 1.0]^T$	$[1.0\ 0.0\ 0.0\ 0.0\ 1.0]^T$
c	0.90	0.90
ρ	0.001	0.001

Table 5.1. Initial conditions and controller parameters for crystallizer moment model.

Note that in both cases, the initial observer states do not match those of the moment model to study the performance of the controller in the presence of initialization errors.

In the first simulation run, we considered achieving the set-point of $\tilde{x}_0 = 0.4$ starting from an initial condition that lies inside the region of guaranteed closed-loop stability (i.e. satisfies the inequality of Eq.5.27). Figure 5.4 shows the closed-loop output (top plot) and manipulated

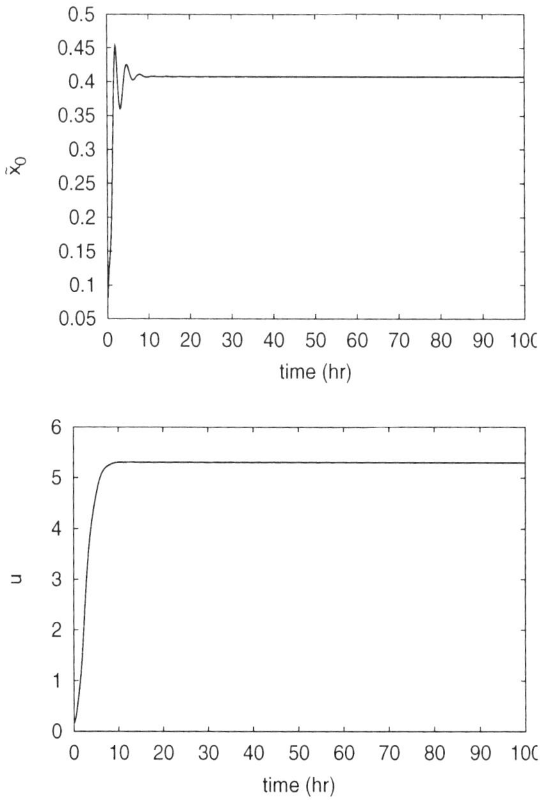

Figure 5.4. Controlled output and manipulated input profiles for the crystallizer moment model under bounded nonlinear controller for $u \in [0, 6]$ and initial condition satisfying Eq.5.29.

input (bottom plot) profiles for this case. It is clear from the figure that, starting from this initial condition, the bounded output feedback controller successfully achieves the requested set-point in the presence of input constraints. In the second simulation run, the same set-point was considered starting from an initial condition that lies outside the region of guaranteed stability (note that the initial condition for the second run given in Table 5.1 does not satisfy Eq.5.27). The closed-loop output and manipulated input profiles for this run are depicted in Figure 5.5, which

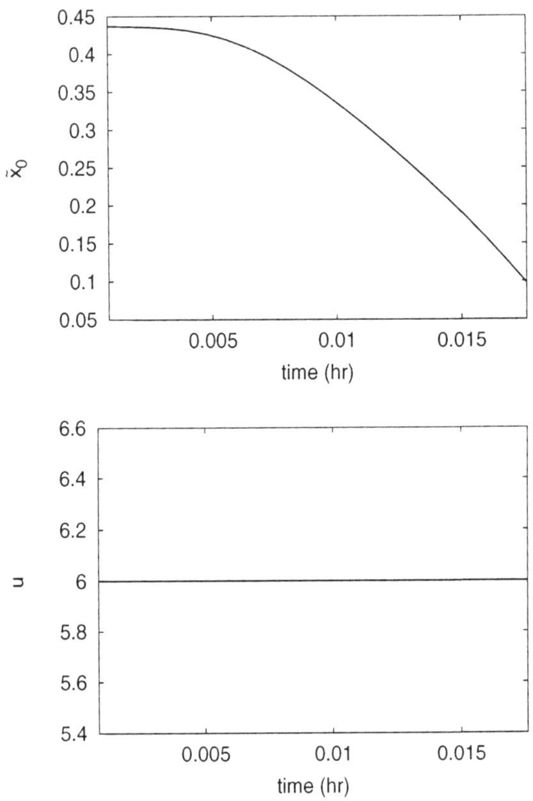

Figure 5.5. Controlled output and manipulated input profiles for the crystallizer moment model under bounded nonlinear controller for $u \in [0, 6]$ and initial condition violating Eq.5.29.

shows that, in this case, the controller fails to achieve the requested set-point in the presence of constraints. These simulation findings verify the prediction of Theorem 5.1 which guarantees the desired properties in the constrained closed-loop ODE system under the bounded output feedback controller of Eq.5.31 for initial conditions that satisfy Eq.5.27.

6.4 Controller Implementation on the Infinite-Dimensional Particulate Process Model

In the previous two subsections, we designed, on the basis of the ODE system of Eq.5.12, a bounded nonlinear output feedback controller with well characterized stability properties in the presence of constraints and illustrated its application to the constrained low-dimensional ODE system of Eq.5.12 which captures the dominant dynamics of the particulate process. In this subsection, we proceed with the final step of our methodology and implement the bounded output feedback controller of Eq.5.28 on the infinite dimensional particulate process model of Eq.5.1. Theorem 5.2 below states precise conditions that guarantee closed-loop stability and asymptotic output tracking in the presence of constraints. The proof of this theorem can be found in appendix 9.C.

Theorem 5.2: *Consider the system of Eq.5.12, for which assumptions 1, 2, and 3 of theorem 1 hold. Consider also the particulate process model of Eq.5.1 under the nonlinear output feedback controller of Eq.5.28. Then, for sufficiently large* N*, there exist positive real numbers* $\delta_n, \bar{\delta}_x, \delta_\omega, \bar{\mu}^*$ *such that if* $\mu \in (0, \bar{\mu}^*]$, $||n(r,0)||_2 < \delta_n$, $|x(0)| < \bar{\delta}_x$, $|\omega(0)| < \delta_\omega$*:*

(1) The closed-loop system (particulate process model and controller of Eq.5.28) is exponentially stable.

(2) $\lim_{t\to\infty} |y_i - v_i| = O(\epsilon(N))$*, where* v_i *is the set-point for the* i*th controlled output and* $\epsilon(N)$ *is a small positive real number that depends on* N *and satisfies* $\lim_{N\to\infty} \epsilon(N) = 0$.

Remark 5.12: Theorem 5.2 establishes that a bounded nonlinear output feedback controller which guarantees exponential stability and output tracking in the constrained finite-dimensional closed-loop system (Eqs.5.12-5.28), continues to enforce the same properties locally in the constrained infinite-dimensional closed-loop system (Eqs.5.1-5.28). This result is intuitively expected because for sufficiently large N: a) the dynamics of the modes of the particulate process model which are not taken into account in the controller design (i.e., not included in the ODE model of Eq.5.12) are locally exponentially stable, and b) the control action, $u(t)$, does not influence the dynamics of the modes which are not taken into account in the controller design.

Remark 5.13: It was pointed out in remark 5.3 that the inequality of Eq.5.27 provides an estimate of the set of admissible initial conditions starting from where the constrained finite-dimensional closed-loop system (Eqs.5.12-5.28) is guaranteed to be stable. Note that owing to the infinite-dimensional nature of the particulate process model of Eq.5.1, this inequality cannot be used directly to check the admissible initial conditions for the infinite-dimensional system (this is because the amplitude of the residual modes not included in the controller design may not, in general, be negligible). Furthermore, the local nature of the result of theorem 5.2 implies that the initial conditions, for the infinite-dimensional system, must be selected sufficiently small to guarantee exponential stability of the closed-loop system. However, the inequality of Eq.5.27 continues to provide a useful guide for the selection of the admissible initial observer states that guarantee stability of the constrained infinite-dimensional closed-loop system (Eqs.5.1-5.28). To see why this is the case, recall from theorem 5.1 that selection of the initial observer states within the largest invariant subset of the region described by Eq.5.27 guarantees stability of the constrained finite-dimensional closed-loop system (and consequently local exponential stability of the infinite-dimensional system according to theorem 5.2). Guided by this result, we conclude that, in order to guarantee stability of the constrained infinite-dimensional closed-loop system, the initial observer states should, at a minimum, not be chosen outside this region. The region of Eq.5.27 therefore provides a reasonable initial guess for where to initiate the observer.

7. Application to a Continuous Crystallizer with Input Constraints

In this section, the proposed nonlinear control method is used to stabilize the continuous crystallizer introduced in subsection 2.2 in the presence of input constraints. Motivated by the fact that the crystallizer with the crystal size distribution as controlled output and the solute feed concentration as manipulated input is an approximately controllable system (see Semino and Ray, 1995a for a rigorous controllability analysis), we consider the control problem of manipulating the solute feed concentration to achieve a crystal size distribution with desired mass, i.e., the controlled output is defined as:

$$y(\bar{t}) \quad = \quad 8\pi\sigma^3 \int_0^\infty n(r,\bar{t})dr = \tilde{x}_0; \qquad (5.33)$$

The reader may refer to Lei et al., 1971; Randolph et al., 1987; Eaton and Rawlings, 1990; Rawlings et al., 1993 for the use of other manipulated variables including fines destruction rate and crystallizer temperature for the stabilization of crystallizers (note that the proposed control method can be used for the synthesis of nonlinear controllers when such manipulated inputs are considered).

Following the proposed methodology outlined in section 5.3, the design of the necessary bounded nonlinear output feedback controller is carried out on the basis of the low-dimensional ODE model (moment model) of Eq.5.16 which captures the dominant dynamics of the crystallizer. The controller design procedure as well as the explicit controller formula are given in section 6.3 of this manuscript where the controller was implemented first on the constrained ODE moment model. In this section, we implement the same controller on the constrained infinite-dimensional crystallizer model of Eq.5.4. The practical implementation of the nonlinear controllers of Eq.5.31 requires on-line measurements of the controlled output, $\tilde{x}_0$; in practice, such measurements can be obtained by using, for example, light scattering (see Bohren and Huffman, 1983; Rawlings et al., 1993 for details).

Several simulation runs were performed to evaluate the performance, robustness, and constraint-handling properties of the bounded nonlinear controller of Eq.5.31 and compare them with those of a PI controller. The values of the nonlinear controller parameters and the PI controller parameters K_c, τ_I, which were used in the simulations, are given in Table 5.2 (K_c, τ_I were computed through extensive trial and error). In all the

	Nominal conditions	with parametric uncertainty
K_c	0.5	0.5
τ_I	1.5	1.0
c	0.90	0.90
ρ	0.001	0.001
τ_i'	–	1.0

Table 5.2. Controller parameters for infinite-dimensional crystallizer model.

simulation runs, the initial condition:

$$n(r, 0) = 0.0, \; c(0) = 990.0 \; kg/m^3$$

was used for the process model of Eq.5.4 and the finite difference method with 1000 discretization points was used for its simulation. The initial conditions for the dynamic system included in the controller of Eq.5.31 were set to be: $\omega_0 = 0.047$, $\omega_1 = 0.028$, $\omega_2 = 0.017$, $\omega_3 = 0.01$ and $\omega_4 = 0.5996$ (note that they do not correspond to the initial conditions used for the distributed parameter model in order to study the performance of the controller in the presence of significant initialization errors).

In the first set of simulation runs, the set-point tracking capability of the nonlinear controller in the presence of input constraints ($0 \leq u \leq 6$) was evaluated under nominal conditions for a 0.4 increase in the value of the set-point ($v = 0.4$). Figure 5.6 shows the closed-loop output (top plot) and manipulated input (middle plot) profiles obtained by using the bounded nonlinear controller (solid lines) of Eq.5.31. For the sake of comparison, the corresponding profiles under PI control are also included (dashed lines). Clearly, the bounded nonlinear controller drives the controlled output to its new set-point value in a significantly shorter

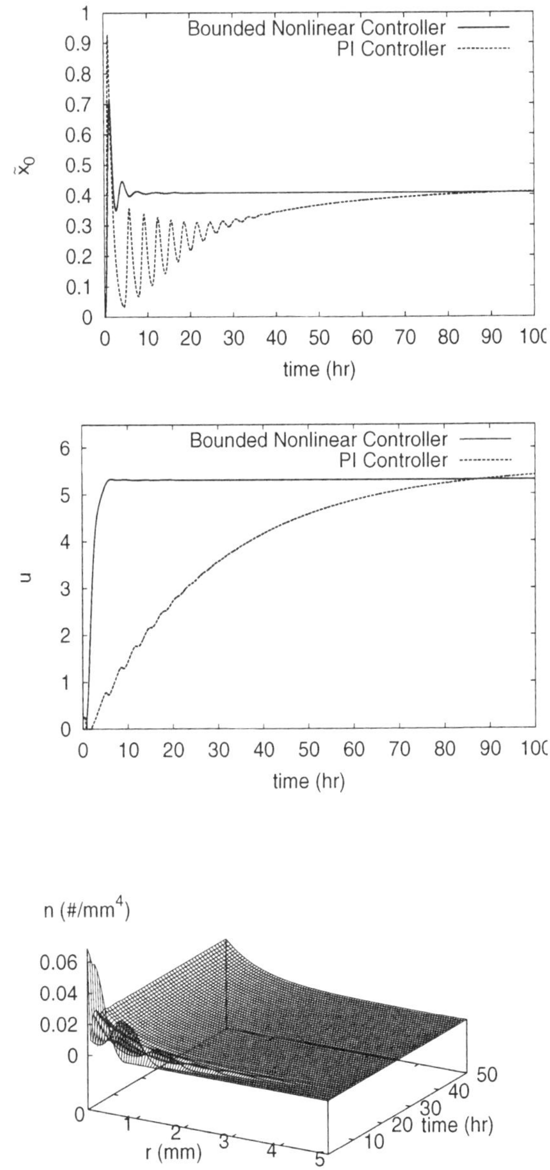

Figure 5.6. Controlled output, manipulated input, and crystal size distribution profiles for the crystallizer process model of Eq.5 under bounded nonlinear controller (solid line) and PI controller (dashed line) for 0.4 increase in the set-point and $u \in [0, 6]$.

time than the one required by the PI controller (note that both controlled outputs exhibit the same overshoot). Note also the superior transient behavior of the closed-loop output under the bounded nonlinear controller compared to the oscillatory response obtained under PI control. For the same simulation run, the evolution of the closed-loop profile of the crystal size distribution is shown in Figure 5.6 (bottom plot.) An exponentially-decaying crystal size distribution is obtained at the steady-state.

Next, the robustness properties of the bounded nonlinear controller in the presence of parametric uncertainties were investigated, for a 0.4 increase in the value of the set-point. To ensure offsetless tracking in the presence of constant uncertainty in process parameters, the bounded nonlinear controller of Eq.5.31 was complemented with integral action (i.e., the term $v - \tilde{h}(\omega)$ was substituted by $v - y + \frac{1}{\tau_i'}\xi$, where $\dot{\xi} = v - y$, $\xi(0) = 0$ and τ_i' is the integral time constant). Figure 5.7 shows the closed-loop output (top plot), manipulated input (middle plot), and evolution of the crystal size distribution (bottom plot) profiles under the bounded nonlinear controller (solid lines) in the presence of 5% error in both F and τ. The corresponding output and input profiles under PI control are also included (dashed lines). We observe that the bounded nonlinear controller exhibits very good robustness properties, driving the output quickly to its new set-point.

Finally, we tested the robustness of the bounded nonlinear controller in the presence of unmodeled actuator and sensor dynamics. To account for the actuator and sensor dynamics, the process model of Eq.5.4 was augmented with the dynamical system $\epsilon_a \dot{z}_1 = -z_1 + z_2$, $\epsilon_a \dot{z}_2 = -z_2 + u$ and the dynamical system $\epsilon_s \dot{z}_3 = -z_3 + z_4$, $\epsilon_s \dot{z}_4 = -z_4 + y$, where $z_1, z_2 \in \mathbb{R}$ are the actuator states, $z_3, z_4 \in \mathbb{R}$ are the sensor states, z_1 is the actuator output, z_3 is the sensor output, and ϵ_a, ϵ_s are small parameters characterizing how fast the actuator and sensor dynamics are, respectively. In this case, the bounded nonlinear controller was also found to be robust with respect to unmodeled dynamics for $\epsilon_a = \epsilon_s = 0.05$. The corresponding closed-loop output (top plot), manipulated input (middle plot), and evolution of crystal size distribution (bottom plot) profiles are depicted in Figure 5.8.

8. Conclusions

In this chapter, we considered spatially-homogeneous particulate processes with input constraints and developed a general methodology for the analysis and control of such processes. Initially, the method of weighted residuals was used to construct finite-dimensional approxima-

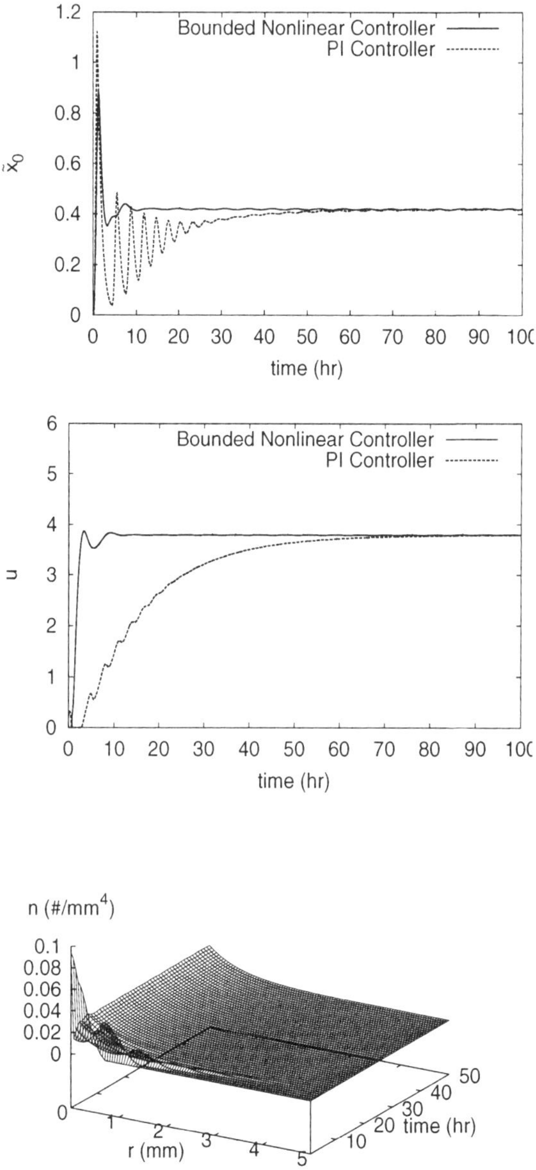

Figure 5.7. Controlled output, manipulated input, and crystal size distribution evolution profiles for the crystallizer process model of Eq.5 under bounded nonlinear controller for 0.4 increase in the set-point in the presence of 5% modeling error in both F and τ and $u \in [0, 6]$.

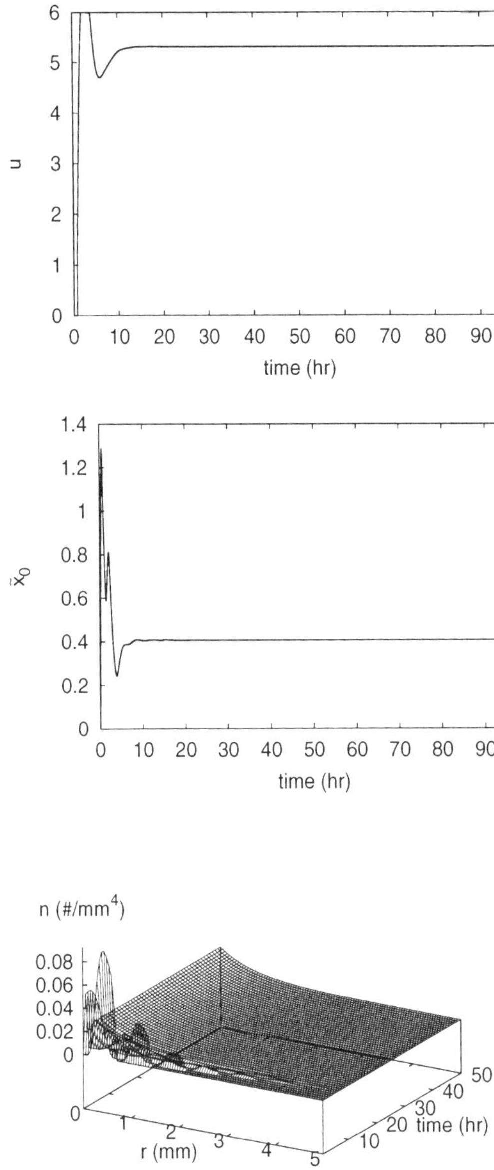

Figure 5.8. Controlled output, manipulated input, and crystal size distribution evolution profiles for the crystallizer process model of Eq.5 under bounded nonlinear controller for 0.4 increase in the set-point in the presence of actuator and sensor unmodeled dynamics and $u \in [0, 6]$.

tions of the particulate process (population balance) model. These ODE systems were then used to identify the set of feasible control objectives (set-points) that can be achieved in the presence of constraints. This information together with the derived ODE systems was then used as the basis for the synthesis of practically-implementable nonlinear bounded output feedback controllers that enforce exponential stability in the closed-loop system and achieve particle size distributions with desired characteristics, in the presence of active input constraints. Precise closed-loop stability conditions were given and controller implementation issues were addressed. The proposed methodology was successfully applied to a continuous crystallizer, which exhibits open-loop unstable (oscillatory) behavior, and shown to cope effectively with the problem of constraints.

Chapter 6

NONLINEAR CONTROL OF SPATIALLY-DISTRIBUTED AEROSOL PROCESSES

1. Introduction

Aerosol processes are increasingly being used for the large scale production of nano- and micron-sized particles. These processes have largely replaced other processes which involve multiple steps of wet chemistry, due to the direct gas phase chemical reaction of precursor vapor and the ease of separation of the particulate products from the gas. Aerosol products, such as TiO_2, B_4C, find widespread use as pigments, reinforcing agents, ceramic powders, optical fibers, carbon blacks and semiconductor materials. Numerous experimental studies have suggested that aerosol growth occurs in stages, beginning with the gas phase chemical reaction of the reactants to produce monomers or molecules of the condensable species (Friedlander, 1977; Friedlander, 1983). The monomers form unstable clusters, which grow further by monomer condensation. Beyond a critical cluster size, nucleation of stable aerosol particles occurs. These particles grow further, mainly by coagulation (condensation and surface reaction are some other growth mechanisms). The coagulation rate, which is affected additively by Brownian and turbulent shear forces, has a strong effect on particle size and morphology.

The dynamic models of aerosol processes are typically obtained from the application of population, material and energy balances, and consist of nonlinear partial integro-differential equation systems (where the independent variables are time, space and one or more internal particle coordinates, such as particle volume, shape). Nonlinearities usually arise from complex reaction, nucleation, condensation and coagulation rates and their nonlinear dependence on temperature. The complex nature of aerosol process models has motivated an extensive research activity on the development of numerical methods for the accurate computation of their solution. Examples of solution methods include the method of self-preserving distributions (e.g., Ramabhadran and Seinfield, 1975; Friedlander, 1977), the method of weighted residuals (Ramkrishna, 1985; Gelbard and Seinfeld, 1978), the sectional method (e.g., Gelbard et al., 1980; Landgrebe and Pratsinis, 1990), and discretization via fixed/moving pivot techniques (e.g., Kumar and Ramkrishna, 1996a; Kumar and Ramkrishna, 1996b). However, even though the above methods can produce accurate solutions of aerosol process models, they are not suitable for the derivation of approximate models which can be used for the synthesis of practically-implementable feedback controllers. Fortunately, the aerosol size distributions obtained by many industrial reactors can be adequately described by lognormal functions, and thus, the dominant dynamic behavior of many aerosol processes can be accurately captured by a model that describes the evolution of the leading moments of the aerosol size distribution (Pratsinis, 1988; William and Loyalka, 1991). Moment models are much simpler to analyze, simulate and utilize for controller design than detailed aerosol process models, while they contain the most important information about the dynamics of the aerosol process.

This chapter proposes a nonlinear feedback control method for spatially-inhomogeneous aerosol processes for which the manipulated inputs, the control objectives and the measurements are distributed in space. Initially, a general nonlinear partial integro-differential equation model which describes aerosol processes with simultaneous chemical reaction, nucleation, condensation, coagulation and convective transport is presented. Then, under the assumption of lognormal aerosol size distribution, the method of moments is employed to reduce the original model into a set of first-order hyperbolic PDEs which accurately describes the spatio-temporal evolution of the three leading moments needed to exactly characterize the aerosol size distribution. This hyperbolic PDE system is then used as the basis for the synthesis of nonlinear distributed output feedback controllers that use process measurements to achieve an aerosol size distribution with desired characteristics (e.g. geometric average particle volume). The controller design problem is addressed by using recent results on nonlinear control of hyperbolic PDE systems (Christofides and Daoutidis, 1996). The performance of the proposed control method is successfully tested through simulations on a typical aerosol process and is shown to be superior to the one of a proportional integral controller. The results of this chapter were first presented in Kalani and Christofides, 1999 (see also Kalani and Christofides, 2002).

2. Preliminaries

2.1 Spatially-distributed Aerosol Process Model

We consider aerosol processes with simultaneous chemical reaction, nucleation, condensation, coagulation and convective transport. A general mathematical model which describes the spatio-temporal evolution of the particle size distribution in such aerosol processes can be obtained from a population balance and consists of the following nonlinear partial integro-differential equation:

$$\begin{aligned}\frac{\partial n}{\partial t} + v_z\frac{\partial n}{\partial z} + \frac{\partial(G(\bar{x}, v, z)n)}{\partial v} - I(v^*)\delta(v - v^*) \\ = \frac{1}{2}\int_0^v \beta(v - \bar{v}, \bar{v}, \bar{x})n(v - \bar{v}, t)n(\bar{v}, t)d\bar{v} \\ -n(v, t)\int_0^\infty \beta(v, \bar{v}, \bar{x})n(\bar{v}, t)d\bar{v}\end{aligned} \tag{6.1}$$

where $n(v, z, t)$ denotes the particle size distribution function, v is the particle volume, t is the time, $z \in [0, L]$ is the spatial coordinate, L is the length scale of the process, v_z is the velocity of the fluid, $G(\bar{x}, v, z), I(v^*), \beta(v - \bar{v}, \bar{v}, \bar{x})$ are nonlinear scalar functions and $\delta(\cdot)$ is the standard Dirac function.

On the other hand, a mathematical model which predicts the spatio-temporal evolution of the concentrations of species and temperature of the gas phase can be obtained from mass and energy balances and has the following form:

$$\frac{\partial \bar{x}}{\partial t} = \bar{A}\frac{\partial \bar{x}}{\partial z} + \bar{f}(\bar{x}) + \bar{g}(\bar{x})b(z)u(t) + \tilde{A}\int_0^\infty a(\eta, v, x)dv \tag{6.2}$$

where $\bar{x}(z, t)$ is an n-dimensional vector of state variables that depend on space and time, $\bar{A}, \tilde{A}$ are constant matrices, $\bar{f}(\bar{x}), \bar{g}(\bar{x}), a(\eta, v, x)$ are nonlinear vector functions, $u(t)$ is the axially distributed manipulated input (e.g. wall temperature) and $b(z)$ is a known function which determines how the control action $u(t)$, is distributed in space (for example, $b(z) = 1$ when $u(t)$ is uniformly distributed in space). The term $\tilde{A}\int_0^\infty a(\eta, v, x)dv$ accounts for mass and heat transfer from the continuous phase to all the particles in the population.

In the population balance of Eq.6.1, the term $\frac{\partial n}{\partial t}$ describes the rate of change of particle concentration in the particle volume interval v, $v + dv$ and in the spatial interval z, $z + dz$, and the term $v_z\frac{\partial n}{\partial z}$ corresponds to convective transport of aerosol particles at velocity v_z. Moreover, the terms $\dfrac{\partial(G(\bar{x},v,z)n)}{\partial v}$ and $I(v^*)\delta(v-v^*)$ account for the loss or gain of particles by condensation at rate G and the formation of new particles of critical volume v^* by nucleation at rate I, respectively. Finally, the gain and loss of particles by Brownian coagulation is captured by the terms:

$$\frac{1}{2}\int_0^v \beta(v-\bar{v},\bar{v},\bar{x})n(v-\bar{v},t)n(\bar{v},t)d\bar{v}, \quad n(v,t)\int_0^\infty \beta(v,\bar{v},\bar{x})n(\bar{v},t)d\bar{v}\ , \tag{6.3}$$

respectively. $G(\bar{x},v,z)$ and β are the condensational growth and collision frequency function, respectively, for which two different expressions are used for the free molecule size and continuum size regimes (Pratsinis, 1988): a) for the free molecule size regime:

$$\begin{aligned} G_{FM}(\bar{x},v,z) &= B_1 v^{\frac{1}{3}}(S-1), \quad B_1=(36\pi)^{\frac{1}{3}}v_1 n_s(k_BT/2\pi m_1)^{\frac{1}{2}} \\ \beta_{FM}(\bar{x},v,z) &= B_2\left(\frac{1}{v}+\frac{1}{\bar{v}}\right)^{\frac{1}{2}}(v^{\frac{1}{3}}+\bar{v}^{\frac{1}{3}})^2, \\ B_2 &= (3/4\pi)^{\frac{1}{6}}(6k_BTv_1/m_1)^{\frac{1}{2}}\ , \end{aligned} \tag{6.4}$$

and b) for the continuum size regime:

$$\begin{aligned} &G_C(\bar{x},v,z)=B_3v^{\frac{1}{3}}(S-1), \quad B_3=(48\pi^2)^{\frac{1}{3}}Dv_1n_s, \\ &D=\lambda(8k_BT/\pi m_1)^{\frac{1}{2}}/3, \quad \beta_C=B_4\left(\frac{C(v)}{v^{\frac{1}{3}}}+\frac{C(\bar{v})}{\bar{v}^{\frac{1}{3}}}\right)\left(v^{\frac{1}{3}}+\bar{v}^{\frac{1}{3}}\right), \\ &B_4=\frac{2k_BT}{3\mu} \end{aligned} \tag{6.5}$$

In Eqs.6.4-6.5, S is the saturation ratio, T is the temperature, D is the condensable vapor diffusivity, λ is the mean free path of the gas, μ is the viscosity of the fluid, n_s is the monomer concentration at saturation ($n_s=P_s/k_BT$, where P_s is the saturation pressure), m_1 is the monomer mass, v_1 is the monomer volume, r is the particle radius, $C(v)=1+B_5\lambda/r$ is the Cunningham correction factor and $B_5=1.257$. Finally, the nucleation rate $I(v^*)$ is assumed to follow the classical Becker-Doring theory and is given by the following expression (Pratsinis, 1988):

$$I \;=\; n_s^2 s_1(k_BT/2\pi m_1)^{\frac{1}{2}}S^2(2/9\pi)^{\frac{1}{3}}\Sigma^{\frac{1}{2}}exp(-k^*lnS/2) \tag{6.6}$$

where s_1 is the monomer surface area and k^* is the number of monomers in the critical size nucleus which is given by:

$$k^* \;=\; \frac{\pi}{6}\left(\frac{4\Sigma}{lnS}\right)^3 \tag{6.7}$$

where $\Sigma = \gamma v_1^{2/3}/k_B T$ and γ is the surface tension.

Remark 6.1: Referring to the system of Eqs.6.1-6.2, several remarks are in order: a) the spatial differential operators are linear; this assumption is valid for spatially-inhomogeneous aerosol processes for which the particle and gas velocities can be considered to be independent of temperature, concentrations and particle size distribution; b) plug-flow is considered and diffusive phenomena in radial and axial directions are neglected; these assumptions are made to simplify the development; c) turbulent coagulation is neglected; this assumption is made to simplify the notation of the theoretical results of the paper and can be readily relaxed; d) the effect of particle curvature on aerosol evaporation rate (Kelvin effect) is neglected; this is done to achieve closure of the integral (moment) aerosol equations (see section 3 below); e) the manipulated input $u(t)$, is a lumped variable (i.e. independent of the spatial coordinate z) and enters the system through Eq.6.2 (mass and energy balance model); this assumption is usually satisfied in most practical applications where the wall temperature is chosen as the manipulated input; and f) the nonlinearities in Eq.6.2 appear in an additive fashion (e.g., complex reaction rates, Arrhenius dependence of reaction rates on temperature).

Remark 6.2: A representative example of an industrial aerosol process which involves coupling of chemical reaction, nucleation, condensation, coagulation and convective transport is the process used for the production of titania powder (see next chapter for modeling and control of this process) through tetra-chloride oxidation. The main feature of this process is that chemical reaction and nucleation take place close to the inlet of the reactor leading to the formation of small particles, and then, coagulation takes place leading to the production of larger particles, while decreasing the total number of particles.

2.2 Methodological Framework for Control of Aerosol Processes

The complexity of the partial integro-differential equation model of Eqs.6.1-6.2 does not allow its direct use for the synthesis of a practically-implementable nonlinear model-based feedback controller for spatially-inhomogeneous aerosol processes. This obstacle, together with the experimental observation that many aerosol size distributions can be adequately approximated by lognormal functions, motivate employing the following methodology for controller design:

1 Initially, the aerosol size distribution is assumed to be described by a lognormal function and the method of moments is applied to the

population balance of Eq.7.2 to compute a hyperbolic PDE system (where the independent variables are time and space) that describes the spatio-temporal behavior of the three leading moments needed to exactly describe the evolution of the lognormal aerosol size distribution.

2 Then, nonlinear geometric control methods (Christofides and Daoutidis, 1996) are applied to the resulting system of hyperbolic PDEs to synthesize nonlinear distributed output feedback controllers that use process measurements at different locations along the length of the process to adjust the manipulated input (typically, wall temperature), in order to achieve an aerosol size distribution with desired characteristics (e.g. geometric average particle volume).

3. Lognormal Aerosol Size Distribution - Moment Model

In this section, we assume that the aerosol size distribution can be adequately represented by a lognormal function and use the method of moments to derive a PDE system which describes the spatio-temporal evolution of the three leading moments of the distribution. The assumption of lognormal aerosol size distribution is motivated by experimental results, as well as from the fact that the parameters of a lognormal function are very frequently used in practice to measure and characterize aerosol size distributions. More specifically, a lognormal aerosol size distribution is described by (Pratsinis, 1988):

$$n(v,t) = \frac{1}{3\sqrt{2\pi} ln\sigma} exp\left(-\frac{ln^2(v/v_g)}{18 ln^2\sigma}\right)\frac{1}{v} \tag{6.8}$$

where v_g is the geometric average particle volume and σ is the standard deviation. Defining the kth moment of the distribution as:

$$M_k(t) = \int_0^\infty v^k n(v,t)dv, \tag{6.9}$$

one can exactly express v_g and σ in terms of the first three moments of the distribution according to the following relations (Brock et al., 1986):

$$v_g = \frac{M_1^2}{M_0^{\frac{3}{2}} M_2^{\frac{1}{2}}}, \quad ln^2\sigma = \frac{1}{9} ln\left(\frac{M_0 M_2}{M_1^2}\right) \tag{6.10}$$

Clearly, the precise characterization of a lognormal aerosol size distribution requires only knowledge of the three leading moments. Therefore, in the remainder of this section, we derive the PDE system that describes the evolution of the first three leading moments of the entire particle size distribution. To this end, we initially compute the moment models for the free molecule size regime and continuum size regime. These models are then used to derive the moment model that describes aerosol dynamics over the entire particle size spectrum by using the harmonic average of the condensation and coagulation rates in the free molecule and continuum regimes.

Free molecule size regime. The PDE system that describes the spatio-temporal evolution of the kth moment of the aerosol size distribution will be computed by substituting Eqs.6.4 into Eq.6.1, multiplying by v^k, and integrating over all particle sizes. Following this procedure, one can show (the detailed calculations are omitted due to space limitations) that the evolution of M_0 (particle concentration), which is only affected by nucleation and coagulation, is governed by:

$$\frac{\partial M_0}{\partial t} = -v_z \frac{\partial M_0}{\partial z} + I - b_0 B_2 (M_{2/3} M_{-1/2} + 2M_{1/3} M_{-1/6} + M_{1/6} M_0) \tag{6.11}$$

where the coefficient b_0 is used for the relationship:

$$\left(\frac{1}{v} + \frac{1}{\bar{v}}\right)^{\frac{1}{2}} = b_0 \left(\frac{1}{v^{\frac{1}{2}}} + \frac{1}{\bar{v}^{\frac{1}{2}}}\right) \tag{6.12}$$

and was computed by the expression $b_0 = 0.633 + 0.092\sigma^2 - 0.022\sigma^3$ (Pratsinis, 1988). The evolution of M_1 (aerosol volume), which is affected by nucleation and condensation, is given by:

$$\frac{\partial M_1}{\partial t} = -v_z \frac{\partial M_1}{\partial z} + Iv^* + B_1(S-1)M_{2/3} \tag{6.13}$$

Finally, the second moment, M_2, depends on nucleation, condensation and coagulation and is described by:

$$\begin{aligned}\frac{\partial M_2}{\partial t} = & -v_z \frac{\partial M_2}{\partial z} + I{v^*}^2 + 2B_1(S-1)M_{5/3} \\ & +2b_2 B_2 (M_{5/3} M_{1/2} + 2M_{4/3} M_{5/6} + M_{7/6} M_1)\end{aligned} \tag{6.14}$$

where b_2 is used as b_0 but for the coagulation kernel of the second moment and is computed by the expression $b_2 = 0.39 + 0.5\sigma - 0.214\sigma^2 + 0.029\sigma^3$ (Pratsinis, 1988).

Continuum size regime. Similar to the case of the free molecule regime, the spatio-temporal evolution of the kth moment of the aerosol size distribution in the continuum size regime is described by a PDE system, which can be obtained by substituting Eqs.6.5 into Eq.6.1, multiplying by v^k, and integrating over all particle sizes. More specifically, the evolution of the zeroth, first and second moments is governed by the following PDEs:

$$\begin{aligned}\frac{\partial M_0}{\partial t} &= -v_z\frac{\partial M_0}{\partial z} + I - B_4[M_0^2 + M_{1/3}M_{-1/3} \\ &\quad + B_5\lambda(4\pi/3)^{\frac{1}{3}}(M_0M_{-1/3} + M_{1/3}M_{-2/3})] ,\end{aligned} \tag{6.15}$$

$$\frac{\partial M_1}{\partial t} = -v_z\frac{\partial M_1}{\partial z} + Iv^* + B_3(S-1)M_{1/3} , \tag{6.16}$$

$$\begin{aligned}\frac{\partial M_2}{\partial t} &= -v_z\frac{\partial M_2}{\partial z} + Iv^{*2} + 2B_3(S-1)M_{4/3} + 2B_4[M_1^2 + M_{4/3}M_{2/3} \\ &\quad + B_5\lambda(4\pi/3)^{\frac{1}{3}}(M_1M_{2/3} + M_{1/3}M_{4/3})]\end{aligned} \tag{6.17}$$

Combining the moment models of Eqs.6.11-6.13-6.14 and Eqs.6.15 -6.16-6.17 by using the harmonic average of the condensation and coagulation rates in the free molecule and continuum regimes, we derive the following dimensionless (see Table 6.1 for the list of dimensionless variables) model for aerosol nucleation, condensation, coagulation and convective transport in the entire particle size spectrum.

Zeroth moment (aerosol concentration):

$$\frac{\partial N}{\partial \theta} = -v_{zl}\frac{\partial N}{\partial \bar{z}} + I' - \xi N^2 \tag{6.18}$$

where:

$$\begin{aligned}\frac{1}{\xi} &= \frac{1}{\xi_{FM}} + \frac{1}{\xi_C}, \\ \xi_{FM} &= r_g'^{\frac{1}{2}}b_0\left[exp\left(\frac{25}{8}ln^2\sigma\right) + 2exp\left(\frac{5}{8}ln^2\sigma\right) + exp\left(\frac{1}{8}ln^2\sigma\right)\right], \\ \xi_C &= K[1 + exp(ln^2\sigma) + B_5(K_{n_1}/r_g')exp\left(\frac{1}{2}ln^2\sigma\right)(1 + exp(2ln^2\sigma))\end{aligned} \tag{6.19}$$

First moment (aerosol volume):

$$\frac{\partial V}{\partial \theta} = -v_{zl}\frac{\partial V}{\partial \bar{z}} + I'k^* + \eta(S-1)N \tag{6.20}$$

where:

$$\begin{aligned}\frac{1}{\eta} &= \frac{1}{\eta_{FM}} + \frac{1}{\eta_C}, \\ \eta_{FM} &= v_g'^{\frac{2}{3}}exp\left(2ln^2\sigma\right), \ \eta_C = \frac{4K_{n_1}}{3}v_g'^{\frac{1}{3}}exp\left(\frac{1}{2}ln^2\sigma\right)\end{aligned} \tag{6.21}$$

Second aerosol moment:

$$\frac{\partial V_2}{\partial \theta} = -v_{zl}\frac{\partial V_2}{\partial \bar{z}} + I'k^{*2} + 2\epsilon(S-1)V + 2\zeta V^2 \tag{6.22}$$

where:

$$\begin{aligned}
\frac{1}{\epsilon} &= \frac{1}{\epsilon_{FM}} + \frac{1}{\epsilon_C}, \quad \frac{1}{\zeta} = \frac{1}{\zeta_{FM}} + \frac{1}{\zeta_C}, \\
\epsilon_{FM} &= {v'_g}^{\frac{2}{3}} exp\left(8ln^2\sigma\right), \; \epsilon_C = \frac{4K_{n_1}}{3}{v'_g}^{\frac{1}{3}} exp\left(\frac{7}{2}ln^2\sigma\right), \\
\zeta_{FM} &= {r'_g}^{\frac{1}{2}} b_2 \\
& exp\left(\frac{3}{2}ln^2\sigma\right)\left[exp\left(\frac{25}{8}ln^2\sigma\right) + 2exp\left(\frac{5}{8}ln^2\sigma\right) + exp\left(\frac{1}{8}ln^2\sigma\right)\right], \\
\zeta_C &= K[1 + exp(ln^2\sigma) \\
& + B_5(K_{n_1}/r'_g)exp\left(-\frac{1}{2}ln^2\sigma\right)(1 + exp(-2ln^2\sigma))
\end{aligned} \tag{6.23}$$

Table 6.1. Dimensionless variables.

$N = M_o/n_s$	Aerosol concentration
$V = M_1/n_s v_1$	Aerosol volume
$V_2 = M_2/n_s {v_1}^2$	Second aerosol moment
$\tau = (2\pi m_1/k_B T)^{1/2}/n_s s_1$	Characteristic time for particle growth
$K = (2k_B T/3\mu)n_s\tau$	Coagulation coefficient
$K_{n_1} = \lambda/r_1$	Monomer Knudsen number
$I' = I/(n_s/\tau)$	Nucleation rate
$R'_r = R_r/(n_s/\tau)$	Reaction rate group
$v'_g = v_g/v_1$	
$r'_g = r_g/r_1$	
$\bar{z} = z/L$	
$v_{zl} = \tau v_z/L$	
$\theta = t/\tau$	

Expressing t and z in Eq.6.2 in dimensionless variables and defining a new vector of state variables $x = [N\ V\ V_2\ \bar{x}]$, the approximate PDE system of Eqs.6.18-6.20-6.22 and the system of Eq.6.2 (which describes the spatio-temporal evolution of concentrations of species and temperature of the continuous phase) yield the following general first-order hyperbolic

PDE system:

$$\frac{\partial x}{\partial \theta} = A\frac{\partial x}{\partial \bar{z}} + f(x) + g(x)b(\bar{z})u(\theta) \tag{6.24}$$

where the specific form of the matrix A and the nonlinear vector functions $f(x), g(x)$ is omitted due to space limitations. The system of Eq.6.24 will be used for controller design in the next section.

Remark 6.3: The computation of the fractional moments in Eqs.6.11-6.13-6.14 and Eqs.6.15 -6.16-6.17 can be performed by expressing the fractional moments in terms of M_0, M_1, M_2 using the relation:

$$\begin{aligned} M_k &= M_0 v_g^k exp\left(\frac{9}{2}k^2 ln^2\sigma\right) \\ &= M_0\left(\frac{M_1^2}{M_0^{\frac{3}{2}}M_2^{\frac{1}{2}}}\right)^k exp\left(\frac{1}{2}k^2 ln\left(\frac{M_0M_2}{M_1^2}\right)\right), \end{aligned} \tag{6.25}$$

which was derived by substituting the lognormal size distribution of Eq.6.8 into Eq.6.9 and using the expressions for v_g and $ln^2\sigma$ of Eq.??.

Remark 6.4: The rate of change of S can be obtained from a monomer balance and is given by:

$$\frac{\partial S}{\partial \theta} = -v_{zl}\frac{\partial S}{\partial \bar{z}} + R_r' - I'k^* - \eta(S-1)N \tag{6.26}$$

4. Nonlinear Feedback Controller Design

In this section, we synthesize nonlinear distributed output feedback controllers for spatially-inhomogeneous aerosol processes described by Eqs.6.1-6.2 that achieve aerosol size distributions with desired characteristics. The controller design problem will be addressed on the basis of the hyperbolic PDE system of Eq.6.24. Such systems are characterized by spatial differential operators whose eigenvalues cluster along vertical or nearly vertical asymptotes in the complex plane, which implies that an infinite number of eigenmodes is required to accurately describe their dynamic behavior. This prohibits the application of modal decomposition techniques to derive ODE models that approximately describe the dynamics of the PDE system and suggests addressing the control problem directly on the basis of the PDE system. The geometric control methodology developed in Christofides and Daoutidis, 1996 will be employed for controller design (the reader may refer to Christofides and Daoutidis, 1996 and Christofides, 2001 for details and proofs of the results).

We initially define a general controlled output, $y(\theta)$, as:

$$y(\theta) = \int_0^1 c(\bar{z})h(x(\bar{z},\theta))d\bar{z} \equiv \mathcal{C}h(x) \tag{6.27}$$

where $h(x(\bar{z},\theta))$ is a nonlinear function of the state of the process (this allows considering controlled outputs that depend in a nonlinear fashion of x, e.g. v_g), $c(\bar{z})$ is a smooth known function which depends on the desired control objective (for example, regulation of v_g in the outlet of an aerosol reactor requires $c(\bar{z}) = \delta(\bar{z}-1)$), and $\mathcal{C}$ is an integral operator. The definition of controlled output allows introducing the concept of characteristic index between the output y and the manipulated input u which will be used in the synthesis of the controller. More specifically, referring to the system of Eq.6.24, the characteristic index of y with respect to u is the smallest integer σ for which (Christofides and Daoutidis, 1996):

$$\mathcal{C}L_g \left(\sum_{j=1}^{n} \frac{\partial x_j}{\partial \bar{z}} L_{a_j} + L_f \right)^{\sigma-1} h(x)b(\bar{z}) \not\equiv 0 \tag{6.28}$$

where a_j denotes the j-th column vector of the matrix $A(x)$, and L_{a_j}, L_f denote the standard Lie derivative notation ($L_f h(x) = \frac{\partial h}{\partial x} f(x)$). From the above definition, it follows that σ depends on the structure of the process (matrix A and functions $f(x)$, $g(x)$, $h(x)$), as well as on the actuator and performance specification functions, $b(\bar{z})$ and $c(\bar{z})$, respectively.

The state feedback control problem is formulated as the one of synthesizing distributed controllers of the general form:

$$u = \mathcal{S}(x) + s(x)y_{sp} \tag{6.29}$$

where $\mathcal{S}(x)$ is a smooth nonlinear operator, $s(x)$ is an invertible matrix of smooth functionals, and y_{sp} is the set-point, that stabilize the closed-loop system and force the following linear input/output response:

$$\gamma_\sigma \frac{d^\sigma y}{d\theta^\sigma} + \cdots + \gamma_1 \frac{dy}{d\theta} + y = y_{sp} \tag{6.30}$$

where $\gamma_1, \gamma_2, \cdots, \gamma_\sigma$ are adjustable parameters which can be chosen to guarantee input/output stability in the closed-loop system.

This controller synthesis problem leads to the following nonlinear distributed state feedback controller (Christofides and Daoutidis, 1996):

$$
\begin{aligned}
u \;=\; & \left[\mathcal{C}\gamma_\sigma L_g\left(\sum_{j=1}^{n}\frac{\partial x_j}{\partial \bar{z}}L_{a_j}+L_f\right)^{\sigma-1}h(x)b(\bar{z})\right]^{-1} \\
& \left\{y_{sp}-\mathcal{C}h(x)-\sum_{\nu=1}^{\sigma}\mathcal{C}\gamma_\nu\left(\sum_{j=1}^{n}\frac{\partial x_j}{\partial \bar{z}}L_{a_j}+L_f\right)^{\nu}h(x)\right\}
\end{aligned}
\tag{6.31}
$$

which enforces stability and the response of Eq.6.30 in the closed-loop system of Eqs.6.24-6.31, provided that the system of Eq.6.24 is minimum-phase (the reader may refer to Christofides and Daoutidis, 1996 for a rigorous definition of the concept of minimum-phase for hyperbolic PDE systems).

The implementation of the distributed nonlinear controller of Eq.6.31 requires knowledge of the state $x(\bar{z},\theta)$ at all positions and times, which may not be available in some practical applications. To overcome this problem, we use the following state observer to estimate the state $x(\bar{z},\theta)$ of the system of Eq.6.24 in space and time:

$$
\frac{\partial \omega}{\partial \theta} \;=\; A\frac{\partial \omega}{\partial \bar{z}}+f(\omega)+g(\omega)b(\bar{z})u+\mathcal{P}(y-\mathcal{C}h(\omega)) \tag{6.32}
$$

where ω denotes the observer state vector and $\mathcal{P}$ is a linear operator, designed on the basis of the linearization of the system of Eq.6.32 so that the system of Eq.6.32 is exponentially stable.

The state observer of Eq.6.32 can be coupled with the state feedback controller of Eq.6.31 to derive a nonlinear distributed output feedback controller of the form:

$$
\begin{aligned}
\frac{\partial \omega}{\partial \theta} \;=\; & A\frac{\partial \omega}{\partial \bar{z}}+f(\omega) \\
& +g(\omega)b(\bar{z})\left[\gamma_\sigma\mathcal{C}L_g\left(\sum_{j=1}^{n}\frac{\partial \omega_j}{\partial \bar{z}}L_{a_j}+L_f\right)^{\sigma-1}h(\omega)b(\bar{z})\right]^{-1} \\
& \left\{y_{sp}-\mathcal{C}h(\omega)-\sum_{\nu=1}^{\sigma}\gamma_\nu\mathcal{C}\left(\sum_{j=1}^{n}\frac{\partial \omega_j}{\partial \bar{z}}L_{a_j}+L_f\right)^{\nu}h(\omega)\right\} \\
& +\mathcal{P}(y-\mathcal{C}h(\omega))
\end{aligned}
$$

$$
\begin{aligned}
u &= \left[\gamma_\sigma \mathcal{C} L_g \left(\sum_{j=1}^{n} \frac{\partial \omega_j}{\partial \bar{z}} L_{a_j} + L_f\right)^{\sigma-1} h(\omega) b(\bar{z})\right]^{-1} \\
&\quad \left\{ y_{sp} - \mathcal{C}h(\omega) - \sum_{\nu=1}^{\sigma} \gamma_\nu \mathcal{C} \left(\sum_{j=1}^{n} \frac{\partial \omega_j}{\partial \bar{z}} L_{a_j} + L_f\right)^{\nu} h(\omega) \right\}
\end{aligned}
\tag{6.33}
$$

that enforces stability and the input/output response of Eq.6.30 in the closed-loop system of Eqs.6.24-6.33.

Remark 6.5: The calculation of the control action from the controllers of Eqs.6.31-6.33 requires algebraic manipulations as well as differentiations and integrations in space, which is expected because of their distributed nature.

Remark 6.6: Note that in the case of imperfect initialization of the observer states (i.e., $\omega(\bar{z}, 0) \neq x(\bar{z}, 0)$), although a slight deterioration of the performance may occur, (i.e., the input/output response of Eq.6.30 will not be exactly imposed in the closed-loop system), the distributed output feedback controller of Eq.6.33 guarantees exponential stability and asymptotic output tracking in the closed-loop system.

Remark 6.7: The exponential stability of the closed-loop system guarantees that in the presence of small errors in process parameters, the states of the closed-loop system will be bounded. Furthermore, one can also couple the nonlinear controller of Eq.6.31 (Eq.6.33) with a proportional integral controller (i.e. substitute $y_{sp} - \mathcal{C}h(x)$ $(y_{sp} - \mathcal{C}h(\omega))$ by $y_{sp} - y + \frac{1}{\tau}\xi_p$, where ξ_p is the integral controller state, which is governed by $\dot{\xi}_p = y_{sp} - y$, $\xi_p(0) = 0$) to ensure asymptotic offsetless output tracking in the closed-loop system, in the presence of constant unknown process parameters and unmeasured disturbance inputs.

Remark 6.8: It can be shown using singular perturbation techniques that the controller of Eq.6.33 possesses a robustness property with respect to unmodeled dynamics provided that they are asymptotically stable and sufficiently fast (i.e., the controller enforces exponential stability and output tracking despite the presence of additional dynamics in the process). This robustness property is of significant practical importance because unmodeled dynamics occur frequently in practice due to actuator and sensor dynamics, as well as other fast process dynamics.

5. Simulation Study

In this section, we describe an application of the proposed nonlinear control method to an aerosol flow reactor used to produce NH_4Cl particles. The following chemical reaction takes place $NH_3 + HCl \rightarrow NH_4Cl$ where NH_3, HCl are the reactant species and NH_4Cl is the monomer species. Under the assumption of lognormal aerosol size distribution, the mathematical model that describes the evolution of the first three moments of the distribution, together with the monomer (NH_4Cl) and reactant (NH_3, HCl) concentrations and reactor temperature takes the form:

$$\begin{aligned}
\frac{\partial N}{\partial \theta} &= -v_{zl}\frac{\partial N}{\partial \bar{z}} + I' - \xi N^2 \\
\frac{\partial V}{\partial \theta} &= -v_{zl}\frac{\partial V}{\partial \bar{z}} + I'k^* + \eta(S-1)N \\
\frac{\partial V_2}{\partial \theta} &= -v_{zl}\frac{\partial V_2}{\partial \bar{z}} + I'k^{*2} + 2\epsilon(S-1)V + 2\zeta V^2 \\
\frac{\partial S}{\partial \theta} &= -v_{zl}\frac{\partial S}{\partial \bar{z}} + C\bar{C}_1\bar{C}_2 - I'k^* - \eta(S-1)N \\
\frac{\partial \bar{C}_1}{\partial \theta} &= -v_{zl}\frac{\partial \bar{C}_1}{\partial \bar{z}} - A_1\bar{C}_1\bar{C}_2 \\
\frac{\partial \bar{C}_2}{\partial \theta} &= -v_{zl}\frac{\partial \bar{C}_2}{\partial \bar{z}} - A_2\bar{C}_1\bar{C}_2 \\
\frac{\partial \bar{T}}{\partial \theta} &= -v_{zl}\frac{\partial \bar{T}}{\partial \bar{z}} + B\bar{C}_1\bar{C}_2\bar{T} + E\bar{T}(\bar{T}_w - \bar{T})
\end{aligned} \tag{6.34}$$

where $\bar{C}_1$ and $\bar{C}_2$ are the dimensionless concentrations of NH_3 and HCl, respectively, $\bar{T}, \bar{T}_w$ are the dimensionless reactor and wall temperatures, respectively, and A_1, A_2, B, C, E are dimensionless quantities. The explicit expressions of A_1, A_2, B, C, E are given in Table 6.2 and the process parameters used in the simulations are given in Table 6.3. Figure 6.1 displays the steady-state profile of the dimensionless total particle concentration, N, as a function of reactor length. As expected, N increases very fast close to the inlet of the reactor (approximately, the first 3% of the reactor) due to a nucleation burst, and then, it slowly decreases in the remaining part of the reactor due to coagulation. Even though coagulation decreases the total number of particles, it leads to the formation of bigger particles, and thus, it increases the geometric average particle volume, v_g. We formulate the control problem as the one of controlling the geometric average particle volume in the outlet of the reactor, $v_g(1, \theta)$, ($v_g(1, \theta)$ is directly related to the geometric average particle diameter, and hence, it is a key product characteristic of industrial aerosol

Table 6.2. Dimensionless variables for the model of Eq.6.34.

$A_1 = \tau k P_o y_{2o}/RT_o$
$A_2 = \tau k P_o y_{1o}/RT_o$
$B = P_o k \tau \Delta H_R y_{1o} y_{2o}/RT_o^2 C_p$
$C = N_{av} k \tau y_{1o} y_{2o} (P_o/RT_o)^2/n_{so}$
$E = 4U RT_o \tau/DC_p P_o$
$\bar{C}_i = y_i/y_{io}\bar{T}$
$\bar{T} = T/T_o$
$\bar{T}_w = T_w/T_o$

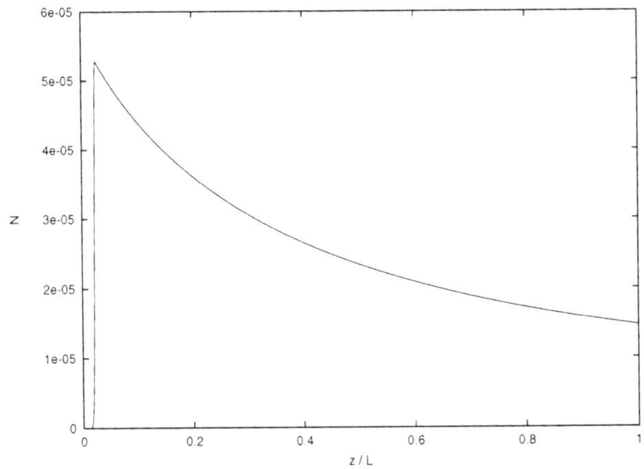

Figure 6.1. Steady-state profile of dimensionless particle concentration.

processes) by manipulating the wall temperature, i.e.:

$$y(\theta) = \mathcal{C}v_g = v_g(1,\theta), \quad u(\theta) = \bar{T}_w(\theta) - \bar{T}_{ws} \tag{6.35}$$

where $\mathcal{C}(\cdot) = \int_0^1 \delta(\bar{z}-1)(\cdot)dz$ and $\bar{T}_{ws} = T_{ws}/T_o = 1$. Since coagulation is the main mechanism that determines the size of the aerosol particles, we focus on controlling the part of the reactor where coagulation occurs. Therefore, the wall temperature is assumed to be equal to its steady-state value in the first 3.5% of the reactor (where nucleation mainly occurs), and it is adjusted by the controller in the remaining part of the reactor (where coagulation takes place).

The model of Eq.6.34 was used as the basis for the synthesis of a nonlinear controller utilizing the proposed control method. For this model, σ was found to be equal to 2 and the necessary controller was synthesized using the formula of Eq.6.31 and is of the form:

$$\begin{aligned} u &= \left[\mathcal{C}\gamma_\sigma L_g\left(\sum_{j=1}^{n}\frac{\partial x_j}{\partial \bar{z}}L_{a_j}+L_f\right)h(x)b(\bar{z})\right]^{-1} \\ &\quad \left\{y_{sp}-\mathcal{C}h(x)-\sum_{\nu=1}^{2}\mathcal{C}\gamma_\nu\left(\sum_{j=1}^{n}\frac{\partial x_j}{\partial \bar{z}}L_{a_j}+L_f\right)^{\nu}h(x)\right\} \end{aligned} \tag{6.36}$$

where $\gamma_1 = 580$ and $\gamma_2 = 1.6 \times 10^5$, to enforce a slightly underdamped response. Two simulation runs were performed to evaluate the set-point tracking capabilities of the nonlinear controller and compare its performance with the one of a proportional integral controller. In both simulation runs, the aerosol reactor was initially assumed to be at steady-state and a 5% increase in the set-point value of $v_g(1,0)$ was imposed at $t = 0\ sec$ (i.e. $y_{sp} = 1.05v_g(1,0)$). Figure 6.2 shows the profile of the controlled output, while Figure 6.3 displays the corresponding profile of

Table 6.3. Process model parameters

$L = 20\ m$
$D = 0.05\ m$
$Re = 2000$
$P_o = 1\ atm$
$T_o = 298\ K$
$y_{1o} = y_{2o} = 40\ ppm$
$U = 10.4\ Jm^{-2}s^{-1}K^{-1}$
$\Delta H_R = 175.7\ KJmol^{-1}$
$C_p = 29.1\ Jmol^{-1}K^{-1}$
$MW_g = 14.0 \times 10^{-3}\ kgmol^{-1}$
$k = 11.4\ m^3mol^{-1}s^{-1}$
$\mu = 3.5 \times 10^{-6}\ kgm^{-1}s^{-1}$
$MW = 53.5 \times 10^{-3}\ kgmol^{-1}$
$logP_s(mmHg) = -4644/T + 0.906logT - 0.00162T + 9.004$
$\gamma = 0.08\ Nm^{-1}$
$v_1 = 5.33 \times 10^{-29}\ m^3$
$R = 8.314\ Jmol^{-1}K^{-1}$
$N_{av} = 6.023 \times 10^{23}\ \#mol^{-1}$
$k_B = 1.38 \times 10^{-23}\ JK^{-1}$

the manipulated input. The proposed nonlinear controller regulates successfully, $v_g(1,\theta)$, to its new set-point value. For the sake of comparison,

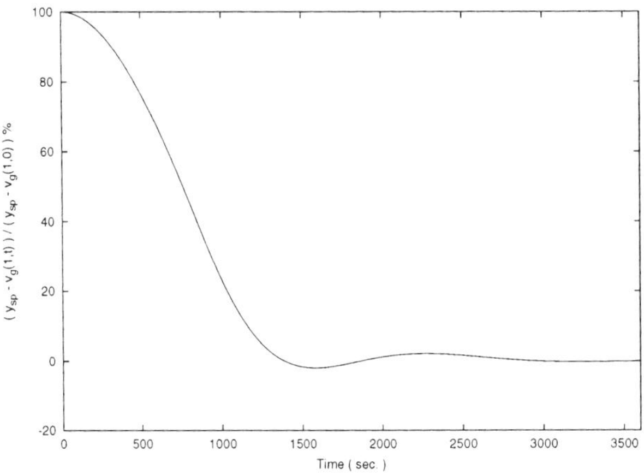

Figure 6.2. Closed-loop profile of v_g in the outlet of the reactor (nonlinear controller).

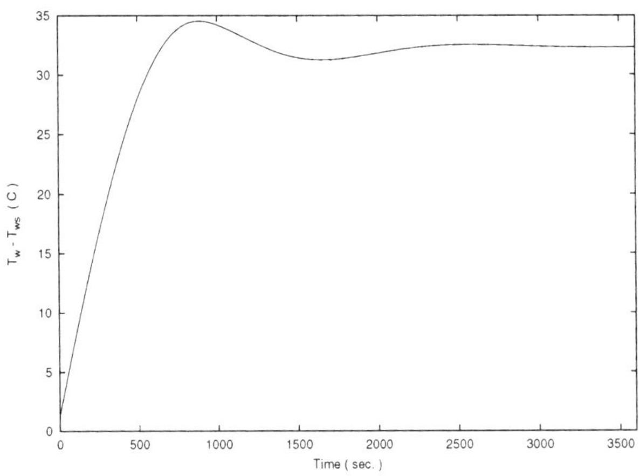

Figure 6.3. Manipulated input profile (nonlinear controller).

we also implemented on the process a proportional integral controller of the form:

$$\begin{aligned} \dot{\xi}_p &= y_{sp} - y, \ \xi_p(0) = 0 \\ u &= K(y_{sp} - y) + \frac{1}{\tau}\xi_p \end{aligned} \tag{6.37}$$

with $K = 3 \times 10^{15}$ and $\tau = 5 \times 10^{-8}$. The profiles of the controlled output and manipulated input are shown in Figure 6.4 and Figure 6.5, respectively (dashed lines show the profiles for the proportional integral controller; the solid lines show the corresponding profiles for the nonlinear controller). It is clear that the proposed nonlinear controller outperforms the proportional integral controller.

Remark 6.9: Regarding the practical implementation of the nonlinear distributed controller of Eq.6.36, we note that the wall temperature, $\bar{T}_w$ is not manipulated directly, but indirectly through manipulation of the jacket inlet flow rate. To this end, a controller should be designed based on an ODE model that describes the jacket dynamics, that operates in an internal loop to manipulate the jacket inlet flow rate to ensure that the jacket temperature obtains the values computed by the distributed robust controller (Christofides and Daoutidis, 1996). Of course, when such a controller is used, a slight deterioration of the closed-loop response obtained under the assumption that $\bar{T}_w$ can be manipulated directly, will occur.

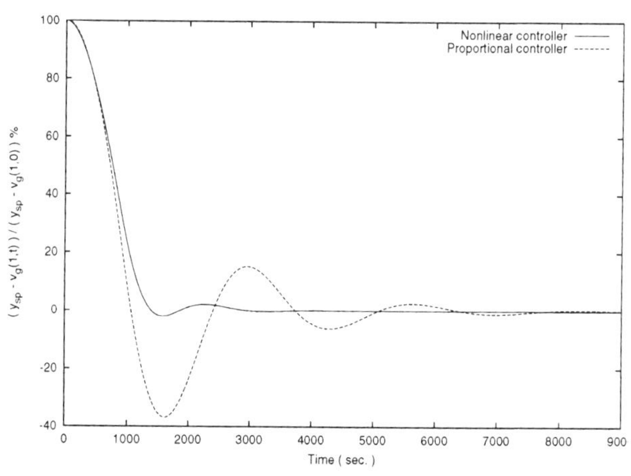

Figure 6.4. Closed-loop profiles of v_g in the outlet of the reactor under proportional integral and nonlinear controllers.

6. Conclusions

In this chapter, we presented a nonlinear feedback control methodology for aerosol processes with simultaneous nucleation, condensation, coagulation and convective transport. Under the assumption that the aerosol size distribution is described by a lognormal function, the method of moments was initially employed to exactly reduce the population balance model into a set of three first-order hyperbolic PDEs which accurately describes the spatio-temporal evolution of the three leading moments of the aerosol size distribution. This system, together with the hyperbolic PDE system that describes the spatio-temporal evolution of the concentrations of species and temperature of the continuous phase, were used as the basis for the synthesis of nonlinear distributed output feedback controllers that use process measurements to attain an aerosol size distribution with desired characteristics. The control method was successfully tested on a typical aerosol process and was shown to outperform a proportional integral control scheme.

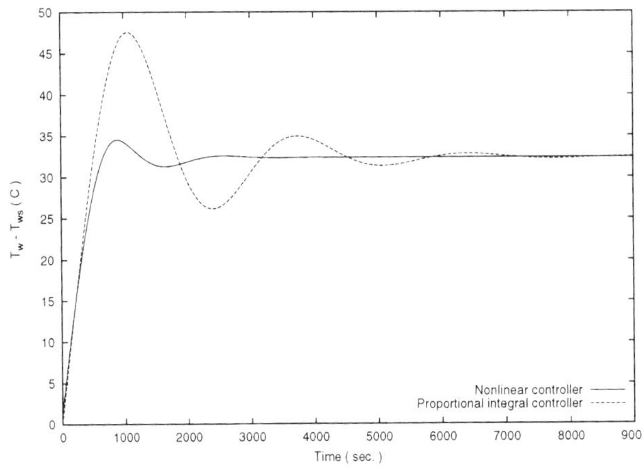

Figure 6.5. Manipulated input profiles for proportional integral and nonlinear controllers.

Chapter 7

CONTROL OF A TITANIA AEROSOL REACTOR

1. Introduction

Titania powder finds a major use as a white pigment in paints and in absorbents, and is commercially produced by two main processes: the gas phase oxidation of titanium tetrachloride, resulting in micron sized particles which are used as white pigments, and the gas phase hydrolysis of titanium tetrachloride, resulting in nanometer sized particles, which are used as UV absorbants. The pigmentary properties of titania, such as the opacity and gloss, depend heavily on the particle size. This fact suggests the design and implementation of advanced model-based feedback control schemes on titania aerosol reactors to produce powders with desired size distributions. The development of mathematical models for aerosol reactors that accurately describe the evolution of key variables that characterize the form of the size distribution is usually addressed through application of population balances to the particulate phase (see Akhtar et al., 1991; Kobata et al., 1991 for modeling of titania aerosol reactors). However, the high complexity of the population balances (systems of nonlinear partial integro-differential equations) renders them inappropriate for process simulation and controller design. Fortunately, the fact that most aerosol size distributions can be adequately described by lognormal functions (William and Loyalka, 1991) allows utilizing the method of moments to derive approximations of the population balances that accurately describe the key properties of the aerosol size distribution; such approximate models are suitable for process simulation and controller design.

This chapter focuses on modeling and control of an aerosol flow reactor used to produce titania powder by gas phase oxidation of titanium tetrachloride; this process is currently used for the production of more than 2 million tons of pigments per year. Initially, a fundamental mathematical model is derived for the process which describes the spatio-temporal evolution of the three leading moments of the aerosol size distribution, as well as the evolution of the concentrations of the species and temperature of the continuous phase. The model accounts for simultaneous chemical reaction, nucleation, Brownian and turbulent coagulation and convective transport and comprises of eight nonlinear first-order hyperbolic partial differential equations (PDEs). Then, the process model is used to synthesize a nonlinear output feedback controller which manipulates the temperature of the reactor wall to achieve an aerosol size distribution in the outlet of the reactor with desired geometric average particle diameter. The performance and robustness of the nonlinear controller is successfully tested through computer simulations. The results of this chapter were first presented in Kalani and Christofides, 2000.

2. Process Description

Titania powder is produced under turbulent and non-isothermal conditions in tubular reactors having very short residence times, high temperatures and relatively low pressures. The premixed and preheated reactants (titanium tetrachloride and oxygen gas) are injected into the reactor where the following exothermic reaction takes place, producing titania monomers and chlorine gas:

$$TiCl_4(g) + O_2(g) \rightarrow TiO_2(s) + 2Cl_2(g) \tag{7.1}$$

The size of a single TiO_2 molecule (monomer) is larger than the thermodynamic critical cluster size. As a result, chemical reaction and nucleation become indistinguishable, thereby implying that the rapid chemical reaction leads to a nucleation burst. The TiO_2 monomers coagulate, causing an increase in the average particle size and a decrease in the particle concentration. In titania aerosol reactors, two forms of coagulation can be distinguished: Brownian coagulation and turbulent coagulation. Brownian coagulation is the dominant mechanism for particle growth when the particle sizes are small, and may occur as free molecule coagulation when the particle size is comparable to the free mean path of the gas, or as continuum coagulation when the carrier gas can be assumed to be homogenous compared to the particulate phase. On the other hand, turbulent coagulation (induced by fluid eddies) becomes dominant when the particle sizes attain macroscopic dimensions. The interplay between Brownian and turbulent coagulation has a strong effect on the shape of the particle size distribution of the titania product (Pratsinis, 1989; Xiong and Pratsinis, 1991). Figure 7.1 presents a schematic diagram of a typical titania aerosol reactor.

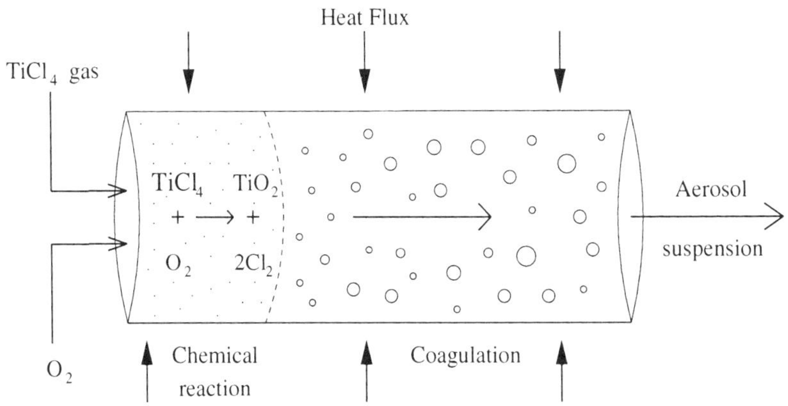

Figure 7.1. Schematic diagram of a titania aerosol reactor.

3. Titania Process Model

In this section, a general model for the titania aerosol process having simultaneous chemical reaction, nucleation, coagulation and convective transport, is presented. The spatio-temporal evolution of the particle size distribution in the process can be obtained from a population balance and is described by the following partial integro-differential equation:

$$\begin{aligned}\frac{\partial n}{\partial t} + \frac{\partial(v_z(z,t)n)}{\partial z} - I(v^*)\delta(v - v^*) \\ = \frac{1}{2}\int_0^v \beta(v - \bar{v}, \bar{v}, \bar{x})n(v - \bar{v}, z, t)n(\bar{v}, z, t)d\bar{v} \\ -n(v, z, t)\int_0^\infty \beta(v, \bar{v}, \bar{x})n(\bar{v}, z, t)d\bar{v}\end{aligned} \tag{7.2}$$

where $n(v, z, t)$ denotes the TiO_2 particle size distribution function, v is the particle volume, t is the time, $z \in [0, L]$ is the spatial coordinate, L is the length of the process, $v_z(z, t)$ is the velocity of the fluid, $I(v^*) and \beta(v - \bar{v}, \bar{v}, \bar{x})$ are nonlinear scalar functions, $\delta(\cdot)$ is the standard Dirac function and $\bar{x}(z, t)$ is an n-dimensional vector of state variables that depend on space and time. Further, the term $\frac{\partial n}{\partial t}$ describes the rate of change of particle concentration in the particle volume interval v, $v + dv$ and in the spatial interval z, $z + dz$, and the term $\frac{\partial(v_z(z,t)n)}{\partial z}$

corresponds to convective transport of aerosol particles. The gain and loss of particles by coagulation is given by the terms:

$$\begin{aligned} &\frac{1}{2}\int_0^v \beta(v-\bar{v},\bar{v},\bar{x})n(v-\bar{v},z,t)n(\bar{v},z,t)d\bar{v}, \text{and} \\ &\qquad n(v,z,t)\int_0^\infty \beta(v,\bar{v},\bar{x})n(\bar{v},z,t)d\bar{v}\ , \end{aligned} \tag{7.3}$$

respectively. $I(v^*)\delta(v-v^*)$ accounts for the nucleation of particles of critical volume v^*. As noted previously, for titania and most other ceramic powders, the nucleation rate is equal to the rate of chemical reaction and the critical volume is the same as the monomer volume. It has been established in Pratsinis et al., 1990 that the high-temperature oxidation of $TiCl_4$ is a first order reaction, thus,

$$I(v_1) \quad = \quad N_{av}kC_1 \tag{7.4}$$

where $k = k_0 exp(-E_{ac}/RT)sec^{-1}$, with $k_0 = 8.26 \times 10^4 sec^{-1}$ and $E_{ac} = 88.8KJmol^{-1}$; N_{av} is the Avogadro's constant $(6.023 \times 10^{23}\#mol^{-1})$, v_1 is the monomer volume and C_1 is the $TiCl_4$ concentration.

The collision frequency function, β is defined as a sum of the Brownian and turbulent shear collision frequency functions (β_B and β_T respectively):

$$\beta \quad = \quad \beta_B + \beta_T \tag{7.5}$$

The Brownian collision frequency function has different expressions for the free molecule size and continuum size regimes (Pratsinis, 1988): a) for the free molecule size regime:

$$\begin{aligned} \beta_{FM}(v,\bar{v},\bar{x}) &= B_1\left(\frac{1}{v}+\frac{1}{\bar{v}}\right)^{\frac{1}{2}}(v^{\frac{1}{3}}+\bar{v}^{\frac{1}{3}})^2, \\ B_1 &= (3/4\pi)^{\frac{1}{6}}(6k_BTv_1/m_1)^{\frac{1}{2}}\ , \end{aligned} \tag{7.6}$$

and b) for the continuum size regime:

$$\beta_C(v,\bar{v},\bar{x}) \quad = \quad B_2\left(\frac{C(v)}{v^{\frac{1}{3}}}+\frac{C(\bar{v})}{\bar{v}^{\frac{1}{3}}}\right)\left(v^{\frac{1}{3}}+\bar{v}^{\frac{1}{3}}\right),\ \ B_2=\frac{2k_BT}{3\mu} \tag{7.7}$$

In Eqs.7.6-7.7, k_B is the Boltzmann's constant $(1.38 \times 10^{-23}JK^{-1})$, T is the temperature, μ is the viscosity of the gas, m_1 is the monomer mass, and $C(v) = 1 + B_3\lambda/r$ is the Cunningham correction factor with r and λ being the particle radius and the mean free path of the gas, respectively, and $B_3 = 1.257$. The gas mean free path, $\lambda = \nu(\pi M_w/2k_BTN_{av})^{1/2}$, where M_w is the molecular weight of the gas. The turbulent shear collision frequency function β_T has the following form (Pratsinis, 1988):

$$\begin{aligned} \beta_T(v, \bar{v}, \bar{x}) &= B_4(v^{\frac{1}{3}} + \bar{v}^{\frac{1}{3}})^3, \quad B_4 = 2.3(3/4\pi)(\epsilon_d/\nu)^{\frac{1}{2}}, \\ \epsilon_d &= 4/D(f/2)^{\frac{3}{2}} v_z^3 \end{aligned} \tag{7.8}$$

where ϵ_d is the turbulent energy dissipation rate, ν is the kinematic viscosity, D is the tube diameter and f is the Fanning friction factor $(0.0791/Re^{\frac{1}{4}})$.

The spatio-temporal evolution of the concentration of species and temperature of the gas phase can be obtained from mass and energy balances and has the following form:

$$\begin{aligned} \frac{\partial C_i}{\partial t} &= -\frac{\partial (v_z C_i)}{\partial z} + \alpha_i k C_i, \quad i = 1, .., 3 \\ \frac{\partial T}{\partial t} &= -\frac{\partial (v_z T)}{\partial z} + (k C_1 \Delta H + U a (T_w - T))/C_{pv} \end{aligned} \tag{7.9}$$

where C_2 and C_3 are O_2 and Cl_2 concentrations, α_1, α_2 and α_3 are -1, -1 and 2 respectively, ΔH is the heat of reaction $(88KJmol^{-1})$, U is the overall heat transfer coefficient $(160Jm_{-2}K_{-1}s_{-2})$, a is heat transfer area per unit volume $(4/D)$, T_w is the wall temperature and C_{pv} is the volumetric heat capacity of the gas phase. Since the chemical reaction rate is very high, most of the $TiCl_4$ is consumed near the reactor inlet. Thus, the physical properties of the gas are calculated by assuming it to be composed of only O_2 and Cl_2. The viscosity of the gas is calculated as follows (Bird et al., 1960):

$$\mu(T) = \sum_{i=1}^{2} \frac{y_i a_i}{y_1 \Phi_{i1} + y_2 \Phi_{i2}} \sqrt{T} \quad kgm^{-1}sec^{-1} \tag{7.10}$$

where y_1 and y_2 are the O_2 and Cl_2 mole fractions. The values of the other parameters used in the equation are given in Table 2.1. The volumetric heat capacity of the gas is evaluated as follows (Kelley, 1960):

$$C_{pv}(T) = 4.18 \sum_{i=1}^{2} C_i \left(d_{i0} + d_{i1} T + \frac{d_{i2}}{T^2} \right) \quad Jm^{-3}K^{-1} \tag{7.11}$$

where the values of the parameters used in the equation are given in Table 7.1.

The fluid flow is assumed to be choked once the gas velocity reaches the acoustic limit. The pressure drop occurring due to frictional losses at the wall (there is no change in the number of moles in the gas phase

Table 7.1. Parameters used in Eqs. 7.10-7.11

$a_1 = 1.73 \times 10^{-6}$
$a_2 = 1.46 \times 10^{-6}$
$\Phi_{11} = 1.0$
$\Phi_{12} = 1.59$
$\Phi_{21} = 0.605$
$\Phi_{22} = 1.0$
$d_{10} = 7.16$
$d_{11} = 10^{-3}$
$d_{12} = -4.0 \times 10^4$
$d_{20} = 8.85$
$d_{21} = 1.6 \times 10^{-4}$
$d_{22} = -6.8 \times 10^4$

due to chemical reaction) is modeled using the following expression (Xiong and Pratsinis, 1991):

$$\begin{aligned} \frac{\partial P}{\partial z} &= -2f\rho v_z^2/D, \qquad v_z < v_{z,sn} \\ &= -2f\rho v_{z,sn}^2/D, \quad v_z \geq v_{z,sn} \end{aligned} \tag{7.12}$$

where P is the pressure, ρ is the gas density and $v_{z,sn}$ is the sonic velocity in the gas medium at a given temperature and is equal to $(\gamma RT/M_w)^{\frac{1}{2}}$, with γ and R being the heat capacity ratio (1.4) and the universal gas constant ($8.314 Jmol^{-1}K^{-1}$) respectively. The fluid dynamic equation thus has the form:

$$\frac{\partial v_z}{\partial t} = -v_z \frac{\partial v_z}{\partial z} + 2fv_z^2/D \tag{7.13}$$

Remark 7.1: Some remarks need to be made for Eqs.7.2 and 7.9: a) the spatial differential operators are nonlinear due to the nonlinearity in the flux terms; b) plug-flow model is considered and diffusive effects in the axial and radial directions are neglected; c) particle deposition on the tube walls is neglected; d) the input variable is the wall temperature T_w, and is lumped (i.e. independent of the spatial coordinate z); e) the nonlinearities in Eq.7.9 are additive (e.g., complex reaction rates, Arrhenius dependence of reaction rates on temperature).

Remark 7.2: In Eq.7.6, the expansion of the expression for β_{FM} yields an unmanageable number of terms necessary for covering a wide range of the particle sizes. To simplify calculations in future sections, the following approximation will be made:

$$\left(\frac{1}{v}+\frac{1}{\bar{v}}\right)^{\frac{1}{2}}=b\left(\frac{1}{v^{\frac{1}{2}}}+\frac{1}{\bar{v}^{\frac{1}{2}}}\right) \tag{7.14}$$

Coefficient b depends on the geometric standard deviation and moments of the particle size distribution (Lee, 1984).

Remark 7.3: In Eq.7.2, it has been assumed that there is instantaneous coalescence between the colliding particles. However, below a temperature of about 1500 K, the coalescence rate becomes comparable to or slower than the coagulation rate and the above assumption no longer holds (Xiong and Pratsinis, 1991). Colliding particles fail to fuse (or coalesce), causing further growth by particle collision to be negligible. Thus, below a temperature of 1500 K, coagulation rate will be assumed as zero.

4. Lognormal Aerosol Moment Model

The complexity of the population balance model (partial integro-differential equation) of Eq.7.2 does not allow its direct use for numerical computation of the size distribution in real-time (a direct discretization of this model through finite-differences or sectional method leads to thousands of ordinary differential equations), as well as for the synthesis of feedback controllers that can be readily implemented in practice. Motivated by the need to circumvent these problems and the experimental observation that titania aerosol volume distributions can be adequately characterized by lognormal (self-preserving) functions, we initially apply the method of moments to Eq.7.2 to compute approximate models that describe the spatio-temporal evolution of the three leading moments of the volume distribution in the free molecule size, continuum size and turbulent shear regimes. Then, under the assumption of lognormal volume distribution, we compute the moment model that describes aerosol dynamics over the entire particle volume spectrum. This approach is based on the fact that moment models provide adequate simplification of the population balance equation by modeling the key average bulk properties of the evolving product.

5. Moment Model

In this section, the partial differential equations describing the spatio-temporal evolution of the three leading moments (i.e., the zeroth, first and second moments) of the volume distribution for the free molecule size, continuum size and turbulent shear regimes, are presented. To this end, we define the kth moment as the particle volume weighted integral of the particle number density function:

$$M_k(z,t) = \int_0^\infty v^k n(v,z,t)dv, \tag{7.15}$$

The computation of the moment equations is done by substituting the appropriate expressions for the nucleation and coagulation rates into Eq.7.2, multiplying by v^k, and integrating over all particle volumes. The detailed derivation of some of the moment equations given below is described in the appendix.

First, the PDE that describes the evolution of the first moment, M_1, which physically corresponds to the aerosol volume, is presented. Since the aerosol volume is independent of the coagulation rate and depends only on the nucleation rate, the PDE that describes the evolution of M_1 is the same for all three regimes, and has the following form:

$$\frac{\partial M_1}{\partial t} = -\frac{\partial(v_z M_1)}{\partial z} + N_{av}kC_1v_1 \tag{7.16}$$

Free molecule size regime. Substituting $\beta = \beta_{FM}$ into Eq.7.2 and integrating over all particle volumes, the spatio-temporal evolution of the zeroth moment, M_0, which represents the aerosol concentration and is affected by nucleation and coagulation, is given by:

$$\begin{aligned}\frac{\partial M_0}{\partial t} &= -\frac{\partial(v_z M_0)}{\partial z} \\ &\quad +N_{av}kC_1 - b_0B_1(M_{2/3}M_{-1/2} + 2M_{1/3}M_{-1/6} + M_{1/6}M_0)\end{aligned} \tag{7.17}$$

where the coefficient b_0 is used as b in Eq.7.14 and is calculated for the zeroth moment. On the other hand, the evolution of the second moment, M_2, which depends both on nucleation and coagulation, is described by:

$$\begin{aligned}\frac{\partial M_2}{\partial t} &= -\frac{\partial(v_z M_2)}{\partial z} \\ &\quad +N_{av}kC_1v_1^2 + 2b_2B_1(M_{5/3}M_{1/2} + 2M_{4/3}M_{5/6} + M_{7/6}M_1)\end{aligned} \tag{7.18}$$

where b_2 is used as b_0 but for the coagulation kernel of the second moment.

Continuum size regime. Similar to the case of the free molecule regime, the PDEs that describe the spatio-temporal evolution of the zeroth and second moments of the aerosol volume distribution in the continuum size regime can be obtained by setting $\beta = \beta_C$ in Eq.7.2, multiplying by v^k (k is equal to 0 and 2, respectively), and integrating over all particle volumes. The resulting PDEs have the following form:

$$\begin{aligned}\frac{\partial M_0}{\partial t} &= -\frac{\partial(v_z M_0)}{\partial z} + N_{av}kC_1 - B_2[M_0^2 + M_{1/3}M_{-1/3} \\ &\quad + B_3\lambda(4\pi/3)^{\frac{1}{3}}(M_0 M_{-1/3} + M_{1/3}M_{-2/3})]\end{aligned} \quad (7.19)$$

$$\begin{aligned}\frac{\partial M_2}{\partial t} &= -\frac{\partial(v_z M_2)}{\partial z} + N_{av}kC_1 v_1^2 + 2B_2[M_1^2 + M_{4/3}M_{2/3} \\ &\quad + B_3\lambda(4\pi/3)^{\frac{1}{3}}(M_1 M_{2/3} + M_{1/3}M_{4/3})]\end{aligned} \quad (7.20)$$

Turbulent shear regime. Finally, following the same approach as in the free molecule and continuum size regimes, we obtain the following PDEs that describe the spatio-temporal evolution of the zeroth and second moments of the aerosol volume distribution in the turbulent shear regime:

$$\frac{\partial M_0}{\partial t} = -\frac{\partial(v_z M_0)}{\partial z} + N_{av}kC_1 - B_4(M_0 M_1 + 3M_{1/3}M_{2/3}) \quad (7.21)$$

$$\frac{\partial M_2}{\partial t} = -\frac{\partial(v_z M_2)}{\partial z} + N_{av}kC_1 v_1^2 + 2B_4(M_1 M_2 + 3M_{4/3}M_{5/3}) \quad (7.22)$$

6. Lognormal Particle Volume Distribution

Studies suggest that the volume distributions of many aerosol products including titania can be adequately approximated by unimodal lognormal functions. In addition, the variables included in lognormal functions, like the geometric average particle volume, are frequently used in industry to characterize aerosol products. Therefore, it is meaningful to represent titania aerosol volume distributions by lognormal functions. Specifically, titania aerosols can be adequately approximated by unimodal lognormal functions of the form (Xiong and Pratsinis, 1991):

$$n(v,z,t) = \frac{1}{3\sqrt{2\pi}ln\sigma} exp\left(-\frac{ln^2(v/v_g)}{18ln^2\sigma}\right)\frac{1}{v} \quad (7.23)$$

where v_g is the geometric average particle volume and σ is the standard deviation. v_g and σ can be expressed in terms of the first three moments of the distribution according to the following relations (Brock et al., 1986):

$$v_g = \frac{M_1^2}{M_0^{\frac{3}{2}} M_2^{\frac{1}{2}}}, \quad ln^2\sigma = \frac{1}{9} ln \left(\frac{M_0 M_2}{M_1^2} \right) \tag{7.24}$$

Furthermore, the moment description of the system can be closed according to the following relation:

$$M_k = M_0 v_g^k exp \left(\frac{9}{2} k^2 ln^2 \sigma \right) \tag{7.25}$$

From Eqs.7.23-7.24, it is clear that lognormal aerosols can be accurately described in terms of the three leading moments of the volume distribution. Therefore, volume distributions of lognormal aerosols can be exactly obtained by simulating the moment model that describes the spatio-temporal evolution of the three leading moments of the distribution over the entire particle volume spectrum. Such a moment model can be obtained as follows. First, the higher-order and fractional moments in Eqs.7.16-7.22 are expressed in terms of the zeroth, first and second moments through the use of Eqs.7.24-7.25. Then, the zeroth and second moment equations for the free molecule size and continuum size regimes are combined to describe the aerosol dynamics over the entire particle volume spectrum by using the harmonic average of the dimensionless coagulation rates in the free molecule size and continuum size regimes (the dimensionless harmonic average coagulation rate reduces to the appropriate limits in the two limiting size regimes and closely approximates the Fuchs-Sutugin approximation in the transition regime (Pratsinis, 1988). Finally, by adding the dimensionless turbulent (shear-induced) coagulation rate to the dimensionless harmonic mean Brownian coagulation rate, the following PDE system that describes the evolution of the three leading moments of the titania aerosol volume distribution over the entire particle volume spectrum is obtained (see Table 7.1 for the list of dimensionless variables):

Zeroth moment:

$$\frac{\partial N}{\partial \theta} = -\phi \frac{\partial (\bar{v}_z N)}{\partial \bar{z}} + k' x_1 - \xi N^2 \tag{7.26}$$

where:

$$
\begin{aligned}
\xi &= \frac{\xi_{FM}\xi_C}{\xi_{FM}+\xi_C}+\xi_T \\
\xi_{FM} &= b_0 {r'_g}^{\frac{1}{2}}\bar{T}^{\frac{1}{2}}\left[exp\left(\frac{25}{8}ln^2\sigma\right)+2exp\left(\frac{5}{8}ln^2\sigma\right)+exp\left(\frac{1}{8}ln^2\sigma\right)\right], \\
\xi_C &= K_1\psi^{-1}\bar{T} \\
&\quad \left[1+exp(ln^2\sigma)+B_3(K_{n_1}/r'_g)exp\left(0.5ln^2\sigma\right)\left(1+exp(2ln^2\sigma)\right)\right], \\
\xi_T &= K_2\eta v'_g exp\left(\frac{9}{2}ln^2\sigma\right)\left[1+3exp\left(-2ln^2\sigma\right)\right]
\end{aligned}
\tag{7.27}
$$

First moment:

$$
\frac{\partial V}{\partial \theta} = -\phi\frac{\partial(\bar{v}_z V)}{\partial \bar{z}}+k'x_1 \tag{7.28}
$$

Second moment:

$$
\frac{\partial V_2}{\partial \theta} = -\phi\frac{\partial(\bar{v}_z V_2)}{\partial \bar{z}}+k'x_1+2\zeta V^2 \tag{7.29}
$$

where:

$$
\begin{aligned}
\zeta &= \frac{\zeta_{FM}\zeta_C}{\zeta_{FM}+\zeta_C}+\zeta_T \\
\zeta_{FM} &= b_2 {r'_g}^{\frac{1}{2}}\bar{T}^{\frac{1}{2}}exp\left(\frac{3}{2}ln^2\sigma\right) \\
&\quad \left[exp\left(\frac{25}{8}ln^2\sigma\right)+2exp\left(\frac{5}{8}ln^2\sigma\right)+exp\left(\frac{1}{8}ln^2\sigma\right)\right], \\
\zeta_C &= K_1\psi^{-1}\bar{T} \\
&\quad \left[1+exp(ln^2\sigma)+B_3(K_{n_1}/r'_g)exp\left(-0.5ln^2\sigma\right)\left(1+exp(-2ln^2\sigma)\right)\right], \\
\zeta_T &= exp\left(9ln^2\sigma\right)\xi_T
\end{aligned}
\tag{7.30}
$$

Finally, the model of the titania aerosol reactor is completed by deriving the dimensionless PDEs which describe the spatio-temporal evolution of the concentration of the species, and the temperature and fluid-dynamics of the gas phase (see Tables 7.1, 7.2 and 7.3 for the list of dimensionless variables, and physical and process parameters):

$$
\frac{\partial \bar{C}_i}{\partial \theta} = -\phi\frac{\partial(\bar{v}_z \bar{C}_i)}{\partial \bar{z}}+\alpha_i k'\bar{C}_i, \quad i=1,..,3 \tag{7.31}
$$

$$\frac{\partial \bar{T}}{\partial \theta} = -\phi \frac{\partial (\bar{v}_z \bar{T})}{\partial \bar{z}} + \left[A k' \bar{C}_1 + B (\bar{T}_w - \bar{T}) \right] \bar{C}_{pv}^{-1} \quad (7.32)$$

$$\frac{\partial \bar{v}_z}{\partial \theta} = -\phi \bar{v}_z \frac{\partial \bar{v}_z}{\partial \bar{z}} + E \bar{v}_z^2 \quad (7.33)$$

Remark 7.4: In Eqs.7.27 and 7.30, $b_0 = 0.633 + 0.092\sigma^2 - 0.022\sigma^3$ and $b_2 = 0.39 + 0.5\sigma - 0.214\sigma^2 + 0.029\sigma^3$, respectively. These polynomial expressions are obtained as functions of σ by regression (Pratsinis, 1988).

7. Analysis of Open-loop System

In this section, the results of several simulation runs of the open-loop (uncontrolled) process are presented, which allow obtaining key insights into the interplay of the various coagulation mechanisms taking place in the process and formulating a meaningful control configuration (i.e., selection of manipulated variable and measurements). The process model of Eqs.7.26-7.28-7.29-7.31-7.32-7.33 was numerically solved by using the method of finite-differences for spatial discretization (the spatial interval was discretized into 200 equispaced subintervals), the computation of the integrals over the entire particle volume spectrum was done by using Wilson's method, and the time-integration was accomplished by using a standard 4th-order Runge-Kutta algorithm. We established through numerous simulations that the result obtained with this discretization scheme are accurate in the sense that further increase in the number of discretization intervals yields almost identical results.

Figure 7.2 displays the steady-state profile of the geometric standard deviation of the particle volume distribution versus the geometric average particle diameter along the axis of the reactor. As discussed previously, the nucleation burst at the beginning of the reactor causes a sudden increase in σ. However, with the rapid completion of the chemical reaction, σ begins to decrease and approaches its asymptotic value under Brownian coagulation (about 1.4). Towards the end of the reactor, shear-induced coagulation becomes important due to the increasing average particle size and extends the higher end of the size distribution, thus causing σ to increase again (Pratsinis, 1989). In this work, the residence time in the reactor is chosen so that the geometric average particle diameter of the aerosol product is the one corresponding to the smallest powder polydispersity.

On the other hand, since the control objective in titania aerosol reactors is to produce titania powder with desired size distribution, we also studied the effect of the wall temperature on the geometric average particle diameter; a variable that is typically used in industry to characterize and compare different aerosol size distributions. Understanding of this effect could lead to a formulation of a meaningful control problem. Figure 7.3 gives the plots of the geometric average particle diameter, d_{pg} with two different values of the wall temperature, $2000K$ and $2400K$. It was assumed that the coagulation rate below $1500K$ is negligible (Xiong and Pratsinis, 1991), hence, there was no increase in d_{pg} in the middle part of the reactor where the process temperature drops below $1500K$ due to Joule cooling. From these results, it is clear that the wall temperature is a variable that has a very significant effect on the geometric average particle diameter, and therefore, it is meaningful choice for manipulated input (see subsection 6.2 below for results of the closed-loop system).

Finally, in order to simplify the presentation of the controller synthesis results in the next section, we rewrite the process model in a vector form. Defining a new vector of state variables $x = [N\ V\ V_2\ \bar{C}_1\ \bar{C}_2\ \bar{C}_3\ \bar{T}\ \bar{v}_z\ ]$, the PDE system of Eqs.7.26-7.28-7.29-7.31-7.32-7.33 can be written in the following general form:

$$\frac{\partial x}{\partial \theta} = A(x)\frac{\partial x}{\partial \bar{z}} + f(x) + g(x)b(\bar{z})u(\theta) \tag{7.34}$$

where $A(x)$ is a matrix which includes elements that may be nonlinear functions of x, $f(x), g(x)$ are nonlinear vector functions, $u(\theta)$ is the manipulated input (wall temperature), and $b(\bar{z})$ is a function that determines how the control action is distributed in space. The specific form of the matrix A and the nonlinear vector functions $f(x), g(x)$ is omitted due to space limitations. The system of Eq.7.34 will be used for controller design in the next section.

8. Nonlinear Control

Motivated by the results of the simulations presented in subsection 7.3., the control problem was formulated as the one of controlling the geometric average particle diameter in the outlet of the reactor by manipulating the wall temperature, i.e.:

$$y(\theta) = \mathcal{C}d_{pg} = d_{pg}(1,\theta), \quad u(\theta) = \bar{T}_w(\theta) - \bar{T}_{ws} \tag{7.35}$$

where $\mathcal{C}(\cdot) = \int_0^1 \delta(\bar{z}-1)(\cdot)d\bar{z}$ and $\bar{T}_{ws} = T_{ws}/T_o = 1$. Since coagulation is the main mechanism that affects the size of the aerosol particles, we take advantage of the fact that coagulation is negligible below 1500 K, to control the coagulation by varying the process temperature. It was verified through open-loop simulations that the process model with the above manipulated input and controlled output is minimum phase.

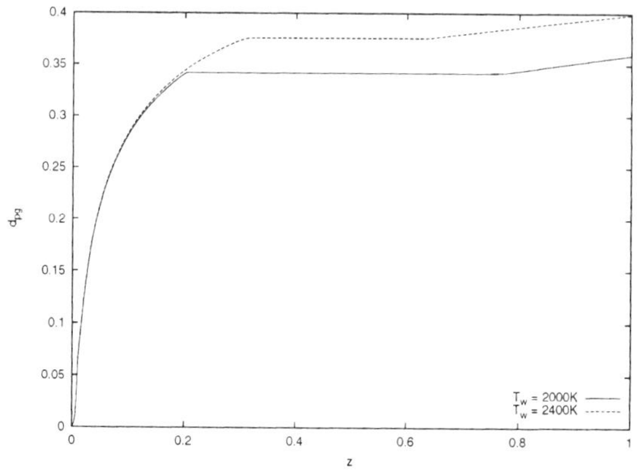

Figure 7.2. Steady-state profile of σ versus $d_{pg}(\mu m)$.

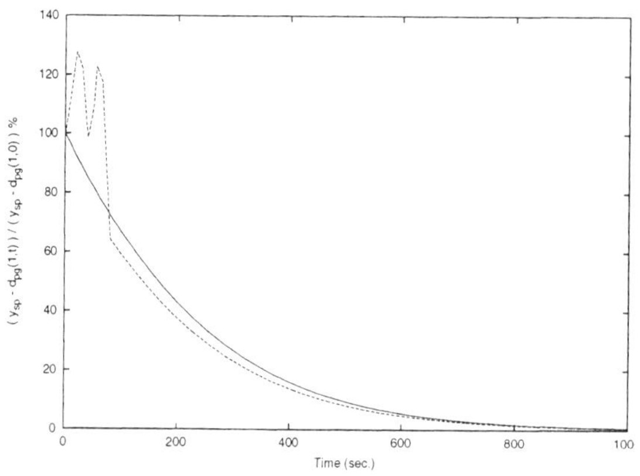

Figure 7.3. Steady-state profile of $d_{pg}(\mu m)$ for two different wall temperature values.

Therefore, the model of Eq.7.34 was used as the basis for the synthesis of a nonlinear controller utilizing the control method described in the previous subsection. For this model, $\bar{\sigma}$ was found to be equal to 2 and the necessary output feedback controller was synthesized using the formula of Eq.6.33 with $y_{sp} - \mathcal{C}h(\omega)$ replaced by $y_{sp} - \mathcal{C}h(x)$. The tuning parameters of the controller were chosen as $\gamma_1 = 50$ and $\gamma_2 = 400$, to enforce a slightly underdamped response in the closed-loop system. Owing to the stability of the open-loop process, the observer gain, $\mathcal{P}$, was set identically equal to zero.

Several simulation runs were performed to evaluate the disturbance rejection and set-point tracking capabilities of the nonlinear controller,

Table 7.2. Dimensionless variables for the model of Eqs. 7.12-7.16

$N = M_0/C_0N_{av}$	Aerosol number concentration
$V = M_1/C_0N_{av}v_1$	Aerosol volume concentration
$V_2 = M_2/C_0N_{av}{v_1}^2$	Second aerosol moment
$\tau_{fm}(T) = (6k_BTr_1v_1/m_1)^{-\frac{1}{2}}/C_0N_{av}$	Characteristic time for free molecule growth
$\tau_c(T) = 3\mu/2k_BTC_0N_{av}$	Characteristic time for continuum growth
$\tau_{sf}(T) = 4\pi(\epsilon_d/\nu)^{-\frac{1}{2}}/6.9v_1C_0N_{av}$	Characteristic time for shear force growth
$\tau_r(T) = 1/k_0exp(-E_{ac}/RT)$	Characteristic time for chemical reaction
$Kn_1 = \lambda/r_1$	Monomer Knudsen number
$\tau_{fm,o} = \tau_{fm}(T_0)$	
$\tau_{c,o} = \tau_c(T_0)$	
$\tau_{sf,o} = \tau_{sf}(T_0)$	
$\tau_{r,o} = \tau_r(T_0)$	
$k' = \tau/\tau_{r,o}$	
$K_1 = \tau_{fm,o}/\tau_{c,o}$	
$K_2 = \tau_{fm,o}/\tau_{sf,o}$	
$\psi = \mu/\mu_o$	
$\eta = (\epsilon_d/\nu)^{\frac{1}{2}}(\epsilon_{d,o}/\nu_o)^{-\frac{1}{2}}$	
$v_g' = v_g/v_1$	
$r_g' = r_g/r_1$	
$\bar{z} = z/L$	
$\bar{v}_z = v_z/v_{z,o}$	
$\phi = v_{z,o}\tau_{fm,o}/L$	
$\theta = t/\tau_{fm,o}$	

as well as its robustness with respect to uncertainty in model parameters and unmodeled actuator dynamics. In all simulation runs, the aerosol reactor was initially assumed to be at steady-state.

In the first set of simulation runs, we tested the ability of the controller to maintain the reactor at the operating steady-state in the presence of three parallel disturbances: one in the inlet feed temperature (ramp decrease upto -5% of steady-state value in 20 sec., constant at -5% until 30 sec., ramp increase to -2.5% in the next 20 sec. and constant thereafter), the second in the inlet mole fraction of $TiCl_4$ (ramp decrease upto -5% of steady-state value in 20 sec., constant at -5% until 30 sec., ramp increase to -2.5% in the next 20 sec. and constant thereafter), and the third in the inlet volumetric flow rate (ramp decrease upto -5% of steady-state value in 20 sec., constant at -5% until 30 sec., ramp increase to -2.5% in the next 20 sec. and constant thereafter). The resulting closed-loop output (top plot - dashed line) and manipulated

Table 7.3. Process model parameters used for the simulation study

$L = 125\ m$	Reactor length
$D = 0.25\ m$	Reactor diameter
$Re = 5 \times 10^5$	Reynolds number
$P_o = 4\ atm$	Inlet pressure
$T_o = 2000\ K$	Inlet temperature
$y_{1o} = 0.4$	Inlet $TiCl_4$ molar fraction
$v_{z,o} = 87.2\ msec^{-1}$	Inlet velocity of process fluid
$C_0 = 24.37\ molm^{-3}$	Inlet concentration of process fluid
$U = 160\ Jm^{-2}s^{-1}K^{-1}$	Overall heat transfer coefficient
$\Delta H_R = 88\ KJmol^{-1}$	Heat of reaction
$C_{pv,o} = 1615.25\ Jm^{-3}K^{-1}$	Inlet volumetric heat capacity of process fluid
$\mu_o = 6.7 \times 10^{-5}\ kgm^{-1}s^{-1}$	Inlet viscosity of process fluid
$v_1 = 3.12 \times 10^{-29}\ m^3$	Monomer volume

Table 7.4. Dimensionless variables for the model of Eqs. 7.17-7.19

$\bar{C}_{pv} = C_{pv}/C_{pv,o}$
$A = C_0 \Delta H / C_{pv,o} T_0$
$B = U a \tau_{fm,o} / C_{pv,o}$
$E = 2 f \tau_{fm,o} \bar{v}_{z,o} / D$
$\bar{C}_i = C_i / C_0$
$\bar{T} = T / T_0$
$\bar{T}_w = T_w / T_0$

input profiles are shown in Figure 7.4; the open-loop output profile (top plot - solid line) under the disturbances is also shown for comparison. One can clearly see that there is an offset (deviation of the d_{pg} at the steady-state from the set-point value) in the case of open-loop operation (uncontrolled process), while when feedback control is implemented on the reactor one can see that this offset has been eliminated. Note that the controller achieves this attenuation of the effect of disturbances without using any measurements of the disturbances. Regarding complete elimination of the effect of disturbances on d_{pg} at the exit of the reactor, we note that this is not possible to be achieved by any control algorithm; this is due to the fact that these disturbances propagate rapidly through the reactor, owing to the very small residence time of the reactor, about

1 *sec*, and affect the process outlet before any control actuator (mechanism with which the wall temperature is adjusted) with realistic limits can react. In the second set of simulation runs, we tested the ability of the controller to maintain the reactor at the operating steady state in the presence of three disturbances in series: one in the inlet feed temperature (ramp decrease upto -5% of steady-state value from 0 to 10 sec., constant from 10 to 20 sec, ramp increase to -2.5% in the next 10 sec. and constant thereafter), the second in the inlet mole fraction of $TiCl_4$ (ramp decrease upto -5% of steady-state value from 20 to 30 sec., constant at -5% from 30 to 40 sec., ramp increase to -2.5% in the next 10 sec. and constant thereafter), and the third in the inlet volumetric flow rate (ramp decrease upto -5% of steady-state value from 40 to 50 sec., constant at -5% from 50 to 60 sec., ramp increase to -2.5% in the next 10 sec. and constant thereafter). Figure 7.5 shows the open-loop output (top plot - solid line), closed-loop output (top plot - dashed line) and manipulated input profiles. Again the controller performs very well regulating the output at the steady-state value, after an initial transient period. From these two simulation runs, it is evident that even though the very small residence time of the reactor and the natural limits on the response of the control actuator do not allow completely eliminating the effect of disturbances on d_{pg} for all times, the use of feedback control helps to attenuate, and eventually eliminate, the effect of disturbances and significantly improve process performance compared to open-loop operation.

In the next simulation run, we tested the set-point tracking capabilities of the controller for a 10% increase in the set-point value of d_{pg} at the exit of the reactor which was imposed at $t = 0$ sec. Figure 7.6 shows the profiles of the controlled output and manipulated input. It is clear that the nonlinear controller smoothly drives the controlled output to the new set-point value, while requesting a smooth control action which is physically realizable (i.e., the rate of change of the wall temperature is less than 2 K per second).

Finally, we tested the robustness of the controller with respect to parametric model uncertainty and unmodeled actuator dynamics. In both simulation runs, a 10% increase in the set-point value of d_{pg} at the exit of the reactor was imposed at $t = 0$ sec. Figure 7.7 shows the closed-loop output and manipulated input profiles in the presence of 5% constant modeling errors in the parameters A, B and E. Again, the nonlinear controller exhibits very good robustness properties, driving the output to its new set point. To test the robustness with respect to unmodeled actuator dynamics, the process model of Eqs. 7.26-7.33 was augmented with the following dynamical system, which represents the actuator dynamics, $\epsilon_z \dot{z}_1 = -z_1 + z_2$, $\epsilon_z \dot{z}_2 = -z_2 + u$, where $z_1, z_2 \in \mathbb{R}$ are the actuator states, z_1 is the actuator output and ϵ_z is a small parameter characterizing how fast are the actuator dynamics. When $\epsilon_z = 0.05$, we found that the nonlinear controller regulates successfully the closed-loop output to its new set-point value exhibiting very good robustness with respect to unmodeled dynamics; the closed-loop output and manipulated input profiles for this simulation run are, as expected, almost identical to the ones shown in Figure 7.6, and therefore, they will be omitted for brevity.

Remark 7.5: We note that we also tested the use of a non-zero gain for the state observer and found that the disturbance rejection properties of the resulting controller are similar to the case where $\mathcal{P} = 0$. The reason is that the control action is constrained by a practical limit that we impose on the rate of change of the wall temperature (about $2K/sec$) which does not allow achieving better rejection of the effect of disturbances.

9. Conclusions

In this work, we studied the modeling and control of an aerosol flow reactor used to produce titania powder by gas phase oxidation of titanium tetrachloride. Initially, a fundamental mathematical model was derived for the process which describes the spatio-temporal evolution of the three leading moments of the aerosol volume distribution, as well as the evolution of the concentrations of the species and temperature of the continuous phase. The model accounts for simultaneous nucleation, Brownian and shear-induced coagulation and convective transport and comprises of eight nonlinear first-order hyperbolic partial differential equations (PDEs). The process model was subsequently used to synthesize a nonlinear output feedback controller, using the method presented in chapter 6, which manipulates the temperature of the reactor wall to achieve an aerosol size distribution in the outlet of the reactor with desired geometric average particle diameter. The performance and robustness of the nonlinear controller were successfully tested through computer simulations.

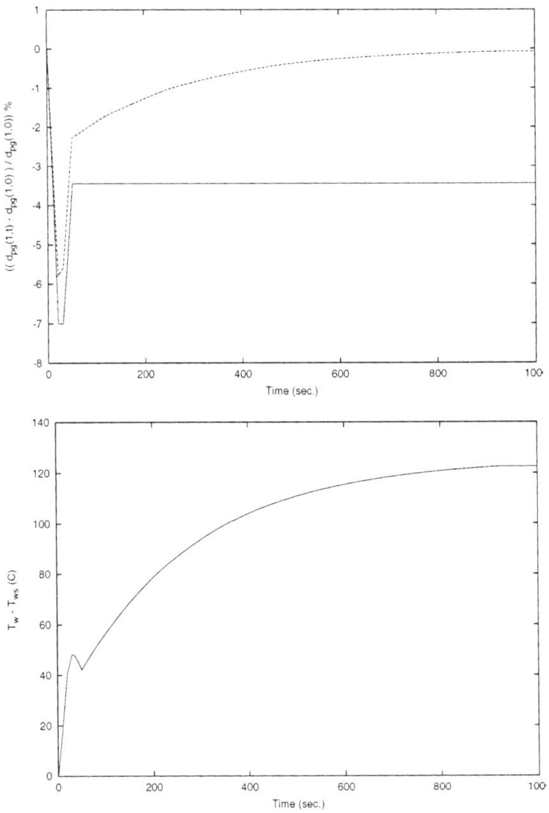

Figure 7.4. Open-loop profile (solid line) and closed-loop profile (dashed line) of $((d_{pg}(1,t) - d_{pg}(1,0))/(d_{pg}(1,0)))\%$ and manipulated input profile under nonlinear control. Parallel disturbances in inlet flow rate, temperature of inlet stream and inlet precursor mole fraction.

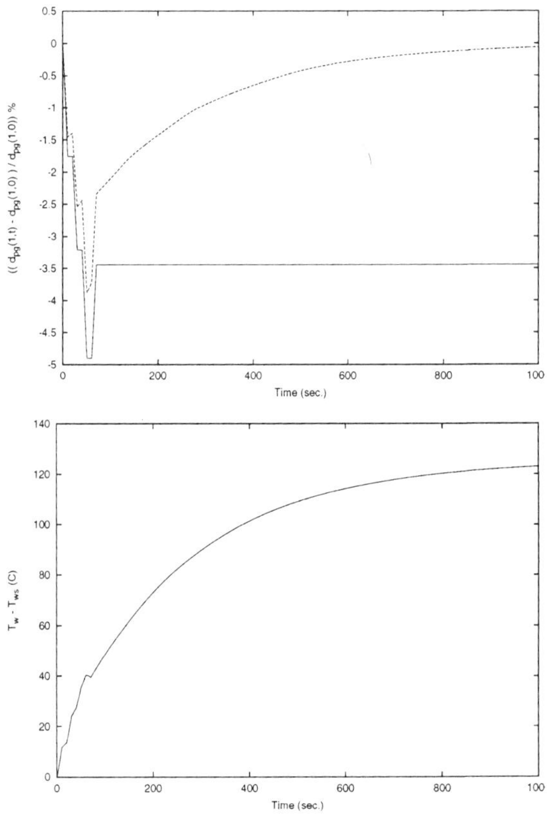

Figure 7.5. Open-loop profile (solid line) and closed-loop profile (dashed line) of $((d_{pg}(1,t) - d_{pg}(1,0))/(d_{pg}(1,0)))\%$ and manipulated input profile under nonlinear control. Disturbances in series in inlet flow rate, temperature of inlet stream and inlet precursor mole fraction.

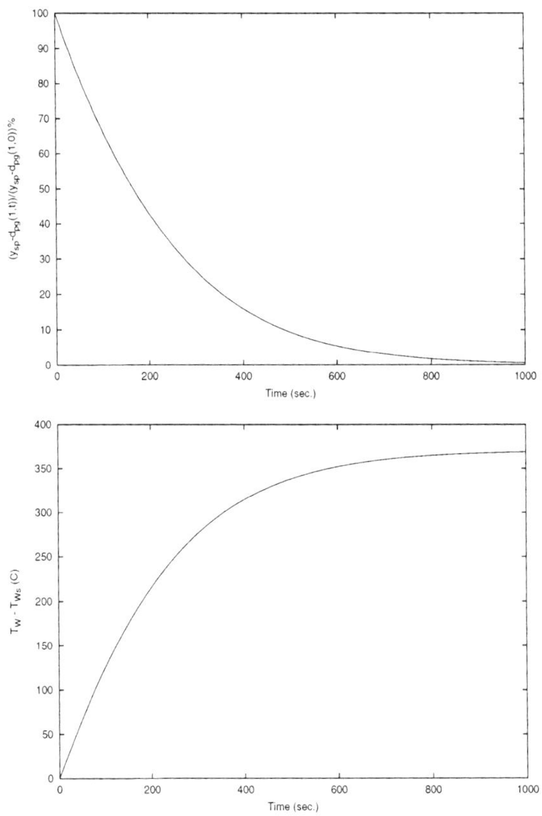

Figure 7.6. Closed-loop profile of $((y_{sp} - d_{pg}(1,t))/(y_{sp} - d_{pg}(1,0)))\%$ and manipulated input profile under nonlinear control. Nominal conditions - set-point change.

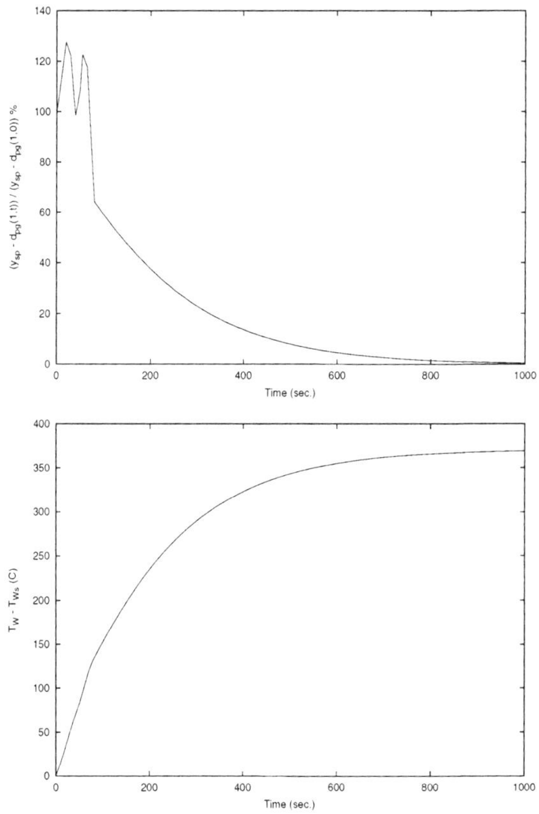

Figure 7.7. Closed-loop profile of $((y_{sp} - d_{pg}(1,t))/(y_{sp} - d_{pg}(1,0)))\%$ and manipulated input profile under nonlinear control. Under parametric uncertainty.

Chapter 8

CONTROL OF AN HVOF THERMAL SPRAY PROCESS

1. Introduction

In this chapter, we focus on the industrially-important high velocity oxygen-fuel (HVOF) thermal spray process and show how the concept of accounting for the effect of powder size distribution in the formulation and solution of the control problem can be used to achieve effective feedback control. The HVOF thermal spray technology is widely used to deposit a large variety of metal and ceramic coatings that modify the surface properties of a base material. Using the thermal energy produced by the combustion of fuel with oxygen to heat and propel the spray particles, it provides a highly efficient way to modify the surface properties of the substrate to extend product life, increase performance and reduce maintenance costs. Recently, there is an increasing interest in the HVOF thermal spray processing of nanostructured coatings, whose grain size is less than about 100 nm (Lau et al., 1998b). This interest has been motivated by several factors, including: (1) the cost-effective production of high-quality nanosize powders; (2) the superior qualities of coatings made with the HVOF process (Cheng et al., 2001a); and (3) the discovery that nanostructured coatings exhibit superior qualities over traditional counterparts (made of materials with micro-sized grains) in several aspects including hardness, strength, ductility and diffusivity (Tellkamp et al., 1997; Lau et al., 1998a; Lau et al., 1999).

Over the last decade, the need to optimally design and operate HVOF thermal spray processes has motivated significant research on the development of fundamental mathematical models to explicitly account for the various physicochemical phenomena and to describe the dynamic behavior of various process components. Specifically, fundamental models have been developed describing the gas dynamics and particle in-flight behavior inside of the gun and in the free jet (e.g., Power et al., 1992; Yang and Eidelman, 1996; Hassan et al., 1998); molten drop deposition, solidification and microstructure development (e.g., Wang et al., 2001); and the relationship between coating microstructure and mechanical properties (e.g., Gourlaouen et al., 2000). In addition, research has been carried out in the integration of the detailed models of the aforementioned components to develop general simulators that describe the behavior of entire thermal spray processes (Yang and Eidelman, 1996; Zagorski and Stadelmaier, 2001).

In order for the HVOF thermal spray process to be widely used in industry for the processing of nanostructured coatings, it is important to achieve excellent real-time process diagnosis and control which will allow obtaining coatings with microstructures that yield the desired properties. Despite the recent progress on the modeling of the various phenomena that affect droplet motion, deposition, solidification and microstructure development in various thermal spray processes, at this stage, there exists no systematic framework for integrated on-line diagnosis and control of the HVOF thermal spray processing which will enable achieving precise regulation of the microstructure and ultimate mechanical and thermal properties of the sprayed materials and coatings. In addition to improving product quality, incorporation of advanced real-time diagnosis and control schemes into thermal spray processes is expected to reduce operational cost and environmental impact, and allow achieving complex (multi-material) coating structures. Since the application of model-based optimization and control techniques to spray casting processes has been reported to lead to significant improvements in their operation and performance (e.g., Annavarapu et al., 1990; Marthur et al., 1991), it is important to develop real-time computer control systems for thermal spray processes by integrating fundamental models that accurately describe the inherent relationships between the coating microstructure and the processing parameters with on-line state-of-the-art diagnostic techniques and control algorithms. Recent efforts in this direction have mainly focused on diagnostics and control of plasma thermal spray by Fincke et al., 2001b; the reader may also refer to the article by Moreau and Leblanc, 2001 for a discussion of process optimization and control issues.

Motivated by the above, we present a formulation of the control problem and a feedback control system for the HVOF thermal spray process which explicitly account for the effect of powder size distribution. Based on the predictions of a fundamental model which is capable of capturing the essential features of the HVOF process, and available experimental data, the control problem is formulated as the one of regulating volume-based averages of the temperature and velocity of the particles at the point of impact on substrate by manipulating the oxygen/fuel ratio and the combustion chamber pressure, respectively. A feedback control system is developed and applied to a detailed model of the process. Closed-loop simulations show that the particle velocity and temperature at the point of impact on substrate reach the desired values in a short time. It is also shown that the proposed formulation of the control problem (which accounts for the effect of powder size distribution) leads to a solution of the control problem that is superior (with respect to the achievement of the desired control objectives) to a solution that assumes a monodisperse powder size distribution. Finally, the proposed control problem formulation and the feedback control system are shown to be robust with respect to disturbances in initial particle velocity and distance between gun exit and substrate. The results of this chapter were first presented in Li and Christofides, 2002.

2. Modeling of the HVOF Process

2.1 Process Description

Figure 8.1 shows the schematic diagram of a commonly used HVOF thermal spray system. It consists of a combustion chamber, a Laval nozzle (also known as convergent/divergent nozzle), and a barrel. High pressure and high temperature combustion gases, which are generated by the combustion of fuel gases (typically propylene, propane and hydrogen etc.) with air in the combustion chamber, are accelerated to supersonic velocity through a Laval nozzle. Outside of the gun, the supersonic free jet adjusts to the ambient pressure by a series of compression and expansion waves, and visible shock diamonds are formed downstream of the barrel exit due to the luminescence of various gases at high temperature. The powder particles are injected axially into the gas stream at the exit of the nozzle, where the pressure is not so high as that in the combustion chamber so that they can easily enter the gas stream. After accelerated and heated in the barrel and in the free jet, the powders impinge on the substrate with high velocity to make coatings.

The HVOF thermal spray process features high gas/particle velocities and relatively low gas/particle temperatures, as compared with plasma spraying (Cheng et al., 2001a). These unique process characteristics contribute to the superior qualities of coatings made with the HVOF process. On the one hand, spraying powders hitting the substrate with high speeds tends to produce coatings with high density and high hardness. On the other hand, relatively low gas temperatures prevent

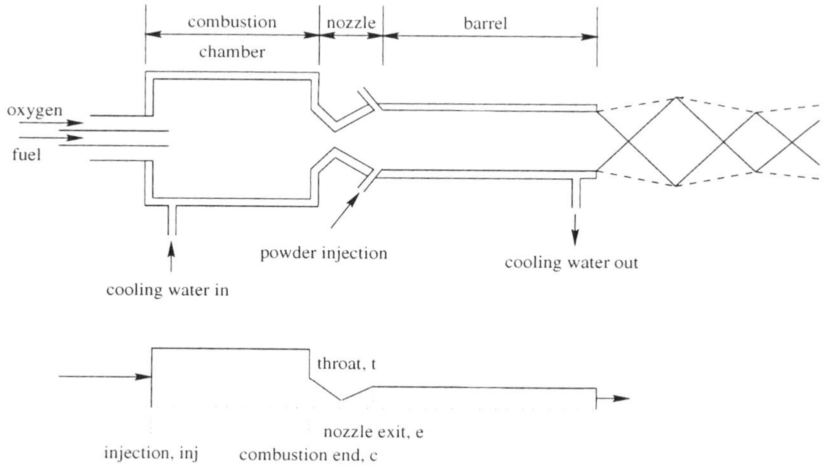

Figure 8.1. Schematic of the HVOF flow field.

particles from being superheated during flight and help to preserve the nanocrystalline structure of powders deposited on the substrate.

From a modeling point of view, HVOF is a complex process which involves combustion, two phase turbulent compressible flow, heat transfer and supersonic/subsonic flow transitions. To be able to develop a fundamental model for this process that is computationally tractable and capable of capturing the main features of the process, we assume in this paper that the presence of particles has a negligible effect on the gas velocity and temperature field. This assumption is standard and reasonably valid because the particle loading, which is defined as the ratio of mass flow rate of particles to that of gases, is typically less than 4% (Yang and Eidelman, 1996). As a consequence, the two-phase problem can be decoupled so that the gas field can be solved first, followed by the simulation of particle infight behavior. To further simplify the analysis, we also make the following standard assumptions: (1) the flame gas obeys the ideal gas law; (2) species reach equilibrium in the combustion chamber and the composition of the combustion products is frozen at the combustion condition along the Laval nozzle and barrel due to the very short residence time of combustion gases in the HVOF gun; (3) the combustion gases behave like a perfect gas during isentropic compression and expansion, and the ratio of specific heat at constant pressure to that at constant volume (c_p/c_v) is nearly a constant; and (4) the friction and cooling water effects along the nozzle and barrel are negligible so that laws of isentropic flow of compressible fluids apply. The assumption of frozen flow is validated in Swank et al., 1994a, which reveals that the combustion products are far from chemical equilibrium, because the values of gas enthalpy, temperature and velocity at the exit of gun barrel as predicted by numerical simulations with the instantaneous equilibrium model are significantly higher than the experimentally measured values (Swank et al., 1994a). Based on the fact that the flow during the Laval nozzle is close to frozen rather than in equilibrium, it is suggested in Swank et al., 1994a that a one dimensional model with chemical equilibrium in the combustion chamber and frozen composition flow during the nozzle and barrel should give a more accurate prediction than the one with instantaneous equilibrium flow. Frozen flow means that the gas residence time is much shorter than the chemical kinetics such that the composition of the combustion gases during the nozzle and the gun barrel is the same as the one in the combustion chamber.

As shown in Figure 8.1, the computational domain includes the combustion chamber, the Laval nozzle and the barrel in the internal field and the supersonic free jet in the external field. There are four important cross sections involved in the internal field: the injection interface (inj), the combustion end (c), the nozzle throat (t) and the nozzle exit (e), which will be discussed later. The modeling procedure is that, for a given combustion chamber pressure (dependent on the total gas flow rate), and fuel/oxygen ratio, we first calculate the chemical composition in the combustion chamber, and then solve the gas properties along the nozzle, in the gun barrel and in the free jet in sequence. The profiles of particle velocity and temperature in the barrel and in the free jet are solved using the momentum transfer equation and the heat transfer equation.

2.2 Modeling of Gas Flow and Thermal Fields

To calculate the equilibrium composition in the combustion chamber, the fuel/oxygen ratio needs to be specified first. In practice, this term is usually expressed by the equivalence ratio:

$$\varphi = \frac{Fuel/Oxidant}{(Fuel/Oxidant)_{st}} \tag{8.1}$$

which is the actual fuel/oxygen ratio divided by the stoichiometric (denoted by the subscript st) fuel/oxygen ratio. Based on this definition, the reaction of a hydrocarbon fuel and air can be represented by a general equation of the form:

$$\varphi C_nH_m + (n + \frac{m}{4})\left[O_2 + \frac{78}{21}N_2 + \frac{1}{21}Ar\right] = \sum_{i\in pr} \xi_i \tag{8.2}$$

where pr represents products and ξ_i is the molar fraction of component i in the combustion products.

Minimization of Gibbs free energy is applied in this paper to calculate the equilibrium composition in the combustion chamber, which shows great advantage than the equilibrium constant method because the specification of a set of reactions a priori, is not necessary. With the assumptions of adiabatic reaction and of ideal gas behavior, the calculation of equilibrium composition in the combustion chamber can be formulated as an optimization problem.

$$\begin{aligned}
min\{G\} &= \sum_{i\in pr} \mu_i\xi_i \\
s.t. & \\
\sum_{i\in pr} a_{ij}\xi_i &= b_j \quad (\forall j \in l) \\
\sum_{i\in re} \xi_i H_i^{\emptyset}(T_0) &= \sum_{j\in pr} \xi_j H_j^{\emptyset}(T_{inj}) \\
\mu_i &= \mu_i(T_{inj})^{\emptyset} + RT_{inj} \ln \frac{P_{inj}\xi_i}{\sum_{i\in pr} \xi_i} \\
& \qquad (\forall i \in pr \ \ with \ \ \xi_i \neq 0) \\
\xi_i &\geq 0 \quad (\forall i \in pr)
\end{aligned} \tag{8.3}$$

where G is the Gibbs energy of the whole system, H_i and μ_i are the enthalpy and chemical potential of species i, respectively; T_0 and T_{inj} represent the room temperature and chamber temperature, respectively; superscript $^{\emptyset}$ represents the standard condition; a_{ij} is the number of element j in species i, b_j the sum of the number of element j in each molecule of reactants and l is the total number of chemical elements involved in the system; R is the gas constant and P is the pressure. The first equation listed in the constraints is the mass balance of each element involved in the system, the second one is the energy balance, and the third one is the definition of chemical potential for ideal gases. Usually, the heat capacity of each species can be correlated as a polynomial of temperature and the enthalpy is derived as the integral of heat capacity with respect to temperature (see Gordon and McBride, 1994 for details).

For the isentropic compressible frozen flow along the Laval nozzle, the governing equations include continuity, momentum balance and energy balance. For example, the momentum balance between the injection interface and the combustion end has the following form:

$$(P + \rho u^2)_{inj} = (P + \rho u^2)_c \tag{8.4}$$

and the corresponding energy balance equation is:

$$\begin{aligned} & \sum_{i \in pr} \xi_i H_i^{\emptyset}(T_{inj}) + \frac{1}{2}(\bar{M}_{pr} \sum_{i \in pr} \xi_i) u_{inj}^2 \\ = & \sum_{i \in pr} \xi_i H_i^{\emptyset}(T_c) + \frac{1}{2}(\bar{M}_{pr} \sum_{i \in pr} \xi_i) u_c^2 \end{aligned} \tag{8.5}$$

where ρ is the gas density, u is the gas velocity and $\bar{M}_{pr}$ is the average molecular weight of the product. In supersonic flow, gas velocity can be expressed by:

$$u_c = \mathbf{M}_c a_c = \mathbf{M}_c \sqrt{\gamma \frac{P_c}{\rho_c}} \tag{8.6}$$

where $\mathbf{M}$ is the mach number which is defined as the ratio of the gas velocity to the local sonic velocity ($a = \sqrt{\gamma P / \rho}$) and γ is the adiabatic constant calculated by $\gamma = \bar{c}_p/(\bar{c}_p - R)$.

During the subsonic/supersonic transition, gases obtain sonic velocity at the throat of the Laval nozzle, or the mach number ($\mathbf{M}_t$) is exactly one. For isentropic flow along the Laval nozzle, the relationship between area ratio and mach number at two cross sections is determined by:

$$\frac{A_2}{A_1} = \frac{\mathbf{M}_1}{\mathbf{M}_2} \left(\frac{1 + \frac{\gamma - 1}{2} \mathbf{M}_2^2}{1 + \frac{\gamma - 1}{2} \mathbf{M}_1^2} \right)^{\frac{\gamma}{2(\gamma + 1)}} \tag{8.7}$$

where A is the cross area perpendicular to the flow direction. For a nozzle with specific configuration, the mach number at the combustion end and the one at the nozzle exit can be solved based on area ratios A_c/A_t and A_e/A_t (see Figure 8.1). Note that the same area ratio corresponds to two mach numbers, with one larger than and the other less than 1, and thus, an appropriate initial guess should be chosen to get the right solution when Newton-Raphson iteration is used.

Combing Eqs.8.4 $\sim$ 8.7 with the ideal gas law, and ignoring u_{inj}, one can solve for the gas properties such as velocity, temperature, pressure, and mach number at the combustion end. Then, the following relationships for isentropic compressible flow, together with Eq.8.7, can be applied to compute these properties at other cross sections along the nozzle:

$$\frac{T_2}{T_1} = \frac{1 + \frac{\gamma - 1}{2}\mathbf{M}_1^2}{1 + \frac{\gamma - 1}{2}\mathbf{M}_2^2} \tag{8.8}$$

$$\frac{P_2}{P_1} = \left(\frac{1 + \frac{\gamma - 1}{2}\mathbf{M}_1^2}{1 + \frac{\gamma - 1}{2}\mathbf{M}_2^2}\right)^{\frac{\gamma}{\gamma - 1}} \tag{8.9}$$

$$\frac{\rho_2}{\rho_1} = \frac{\mathbf{M}_1}{\mathbf{M}_2}\left(\frac{1 + \frac{\gamma - 1}{2}\mathbf{M}_1^2}{1 + \frac{\gamma - 1}{2}\mathbf{M}_2^2}\right)^{\frac{1}{\gamma - 1}} \tag{8.10}$$

The above equations are derived from the aforementioned governing equations for isentropic compressible flow and the reader may refer to the book by Roberson and Crowe, 1997 for a detailed derivation. The mass flow rate through a Laval nozzle, denoted by $\dot{m}$, is determined by

$$\dot{m} = \rho_t u_t A_t = \sqrt{\gamma \rho_t P_t} A_t \tag{8.11}$$

where the subscript t denotes the throat of the nozzle.

For the external flow field, experimental work based on Laser Doppler Velocimeter (LDV) of Tawfik and Zimmerman, 1997 shows that the mean axial velocity and temperature decay of the free supersonic jet in the external gas field can be expressed by the following empirical formulae:

$$\begin{aligned} \frac{u}{u_e} &= 1 \quad (x \leq L_{pc}) \\ \frac{u}{u_e} &= 1 - \exp\left(\frac{0.85}{1 - x/L_{pc}}\right) \quad (x > L_{pc}) \end{aligned} \tag{8.12}$$

and

$$\begin{aligned} \frac{T - T_a}{T_e - T_a} &= 1 \quad (x \leq L_{pc}) \\ \frac{T - T_a}{T_e - T_a} &= 1 - \exp\left(\frac{1.25}{1 - x/L_{pc}}\right) \quad (x > L_{pc}) \end{aligned} \tag{8.13}$$

where x is the axial distance from the exit of the gun barrel and L_{pc} is the potential core length, which can be correlated to a function of the mach number at the exit of gun barrel ($\mathbf{M}_e$) and barrel diameter (D)

$$\frac{L_{pc}}{D} = 3.5 + 1.0\mathbf{M}_e^2 \quad (1.5 \leq \mathbf{M}_e \leq 2.5) \tag{8.14}$$

2.3 Modeling of Particle Motion/Temperature

The particle trajectories and temperature histories in the gas field are computed by the momentum and heat transfer equations. Initially, because the acceleration and deceleration of particles in the moving gas in the HVOF thermal spray system are dominated by the drag force (Pawlowski, 1995), the particle motion can be described by the following two first-order ordinary differential equations:

$$\begin{aligned} m_p \frac{dv_p}{dt} &= \frac{1}{2} C_D \rho_g A_p (v_g - v_p) \mid v_g - v_p \mid, \;\; v_p(0) = 0 \\ \frac{dx_p}{dt} &= v_p, \;\; x_p(0) = 0 \end{aligned} \tag{8.15}$$

where v_p is the particle axial velocity, A_p is the projected area of the particle on the plane perpendicular to the flow direction, C_D is the drag coefficient, and x_p is the particle position, calculated from the exit of the nozzle. Note that one difference between particle motion in a stagnant fluid and particle motion in a moving fluid is that there is an absolute sign of the relative velocity between particle and gas in the latter case, which guarantees that a particle is accelerated if its velocity is less than that of the gas and decelerated otherwise.

To take into consideration the fact that the powders used in the HVOF process are not, in general, spherical (Cheng et al., 2001b), a formula, proposed by Ganser, 1993; Cheng et al., 2001b, for the drag coefficient (C_D) which accounts for the sphericity (ϕ, defined as the ratio of the surface area of a sphere with equivalent volume to the actual surface area of the particle) is used in this paper:

$$\begin{aligned} \frac{C_D}{K_2} &= \frac{24}{ReK_1K_2}\left[1 + 0.1118(ReK_1K_2)^{0.6567}\right] \\ &\quad + \frac{0.4305}{1 + 3305/ReK_1K_2} \end{aligned} \tag{8.16}$$

where K_1 and K_2 are the Stokes' shape factor and Newton's factor, respectively, which can be determined by

$$\begin{aligned} K_1^{-1} &= \frac{1}{3} + \frac{2}{3\sqrt{\phi}} \\ \log(K_2) &= 1.8148(-log\phi)^{0.5743} \end{aligned} \tag{8.17}$$

for isometric shapes. The local Reynolds number (Re) is defined based on the relative velocity

$$Re = \frac{d_p \mid v_g - v_p \mid \rho_g}{\eta_g} \tag{8.18}$$

where η_g is the gas viscosity and d_p is either the particle diameter if the particle is spherical or the equivalent diameter if not.

In the HVOF process, because the Biot number ($Bi = hx/\lambda_p$, h is the heat transfer coefficient, x is a characteristic dimension defined by the ratio of particle volume to its surface area, and λ_p is the thermal conductivity of the particle) is typically less than 0.1 (Cheng et al., 2001b; Yang and Eidelman, 1996), particles are heated with negligible internal resistance and temperature gradients inside the particles can be ignored (Geankoplis, 1993). Consequently, the equation describing the heat transfer between a single particle and the gas reduces to a first-order ordinary differential equation of the form:

$$m_p c_{p_p} \frac{dT_p}{dt} = hA'_p(T_g - T_p), \;\; T_p(0) = T_0 \tag{8.19}$$

where T_p is the particle temperature, m_p is the particle mass, A' is the surface area of a particle and h is the heat transfer coefficient computed by the Ranz-Marshall empirical equation (Bird et al., 1960):

$$\frac{hd_p}{\lambda_g} = Nu = 2 + 0.5 Re^{\frac{1}{2}} Pr^{\frac{1}{3}} \tag{8.20}$$

where the Prandtl number (Pr) is calculated by $Pr = c_{p_g}\eta_g/\rho_g$.

The equations for particle motion and particle temperature are solved by numerical integration. At each step, we integrate Eqs.8.15 and 8.19 with a small enough time step such that the gas velocity, gas temperature, the local Reynolds number, and Prandtl number can all be considered constant over this interval. After one integration step, we update the gas velocity and gas temperature according to the new particle position and then apply the same strategy for the next time step. This methodology was proposed by Crowe and Stock, 1976 and was found to be computationally economical and accurate.

2.4 Powder Size Distribution

The powders used in the HVOF process are polydisperse. Recently, experimental work (e.g., Xu and Lavernia, 2001) has shown that lognormal functions can adequately describe many powders used in the HVOF process. To this end, we use a lognormal function to describe the powder size distribution of the form (Crow and Shimizu, 1988):

$$f(d_p) = \frac{1}{\sqrt{2\pi}\sigma d_p} \exp\left[-\frac{(\ln d_p - \mu)^2}{2\sigma^2}\right] \tag{8.21}$$

where $f(d_p)$ is the size distribution function, μ and σ^2 are two dimensionless parameters corresponding to the mean and the variance of random variable $\ln d_p$, which obeys the normal distribution. For particles that are lognormally distributed, the cumulative volume or weight function is defined as

$$F(d_p) = \frac{\int_0^{d_p} \frac{1}{6}\pi\rho d_p^3 f(d_p) d(d_p)}{\int_0^{\infty} \frac{1}{6}\pi\rho d_p^3 f(d_p) d(d_p)} \tag{8.22}$$

which can be simplified as

$$F(d_p) = \int_{-\infty}^{\frac{\ln d_p - (\mu + 3\sigma^2)}{\sigma}} \frac{1}{\sqrt{2\pi}} e^{-\frac{x^2}{2}} dx \tag{8.23}$$

Once the value of $F(d_p)$ is specified, the upper bound of the above integral can be found from standard integration tables. In experimental powder measurements, d_{10}, d_{50} and d_{90}, which denote the particle sizes corresponding to 10, 50 and 90 % of the cumulative weight function, can be determined using a Microtrac Standard Range Particle Analyzer (see Lau et al., 1998b for experimental results). These measurements can be related to μ and σ according to the following expression:

$$\begin{aligned} \ln d_{10} - (\mu + 3\sigma^2) &= -1.28\sigma \\ \ln d_{50} - (\mu + 3\sigma^2) &= 0 \\ \ln d_{90} - (\mu + 3\sigma^2) &= 1.28\sigma \end{aligned} \tag{8.24}$$

In case there are three sets of data available, parameter estimation can be applied by minimizing the square sum of differences between experimental data and theoretical predictions, i. e.

$$\begin{aligned} \min\{g\} = & [\ln d_{10} - (\mu + 3\sigma^2) + 1.28\sigma]^2 \\ & +[\ln d_{50} - (\mu + 3\sigma^2)]^2 \\ & +[\ln d_{90} - (\mu + 3\sigma^2) - 1.28\sigma]^2 \end{aligned} \tag{8.25}$$

The solution to the above minimization problem yields:

$$\begin{aligned} \mu &= \ln \sqrt[3]{d_{10}d_{50}d_{90}} - 1.831\left(\ln\sqrt{\frac{d_{90}}{d_{10}}}\right)^2 \\ \sigma &= 0.781 \ln\sqrt{\frac{d_{90}}{d_{10}}} \end{aligned} \tag{8.26}$$

Finally, the volume-based or weight-based average of powder properties (PP) can be calculated from the following equation:

$$\overline{PP} = \frac{\int_0^\infty \frac{1}{6}\pi\rho d_p^3 PP(d_p) f(d_p) d(d_p)}{\int_0^\infty \frac{1}{6}\pi\rho d_p^3 f(d_p) d(d_p)} \tag{8.27}$$

3. Simulation Results and Analysis

All the simulations are based on the parameters listed in Table 8.1. Parameters marked with * are average values.

Parameter	*Quantity/Value*	
Fuel	*Propane*	
Oxidant	*Air*	
Powder	*Inconel*	
A_c/A_t	8.0	
A_e/A_t	2.0	
η_g^*	5×10^{-5}	$kg/m \cdot s$
λ_g^*	0.08	$W/m \cdot K$
ρ_p	9000	kg/m^3
c_{p_p}	462	$J/kg \cdot K$
ϕ	1.0	
Barrel Diameter	1.0	cm
Barrel Length	0.2	m
Spray Distance	0.3	m

Table 8.1. Parameters used in the process model.

Table 8.2 shows the gas properties in the internal flow field of the HVOF gun under operating conditions P = 9.0 *bar* and ϕ =1.0. The thermal energy is generated by the reaction of fuel with oxygen in the chamber and then converted into kinetic energy through the nozzle.

	Chamber	*Throat*	*Nozzle Exit*
$P(bar)$	9.000	4.986	1.002
$u(10^3 m/s)$	–	0.871	1.552
$T(10^3 K)$	2.315	2.055	1.473
M	–	1.000	2.092

Table 8.2. Simulation results of gas dynamics in the internal flow field - $P = 9.0$ bar and $\phi = 1.0$.

Figures 8.2 and 8.3 show the gas temperature in the combustion chamber as well as the gas velocity and temperature at the exit of the nozzle under different operating conditions. In Figure 8.2, we fix chamber pressure at 9 *bar* and change the equivalence ratio from 0.8 to 1.2. It is found that in the equivalence ratio range of interest, gas velocity at the exit of the nozzle changes very little compared to the gas temperature. As the equivalence ratio increases from 0.8 to 1.2, gas temperature at the exit of the nozzle increases initially, reaching its maximum value 1477.7 K at equivalence ratio slightly above 1.0, and then decreases. A similar phenomenon is also observed for the gas temperature in the combustion chamber. From these it can be concluded that fuel combustion under near stoichiometry conditions generates more thermal energy and achieves higher gas temperature. In Figure 8.3, we keep equivalence ratio constant at 1.0 and change pressure from 5 *bar* to 12 *bar*. It can be seen that gas temperature in the combustion chamber as well as gas velocity and gas temperature at the exit of the nozzle do not change in the chamber pressure range of interest.

Figures 8.4 and 8.5 display the axial mean velocity and temperature of particles with different sizes along the flow field. It can been seen that all the particles with different sizes are accelerated in the barrel and in some region of the supersonic free jet (where their velocities are less than that of the combustion gas), and are then decelerated as they reach the same velocity as that of the gas. However, fine particles change their velocities more easily than coarser ones because of the less momentum inertia. For example, a particle with a diameter about 5 μm attains velocity as high as about 1300 m/s during flight. However, its velocity decreases rapidly because of entrainment, which explains why very small size particles are not suitable for thermal spraying. For particles with large size, the velocity profiles become nearly flat after they reach the same velocity of the gas, which implies that the drag forces exerted by the gas are too small to change the velocities of particles. A similar

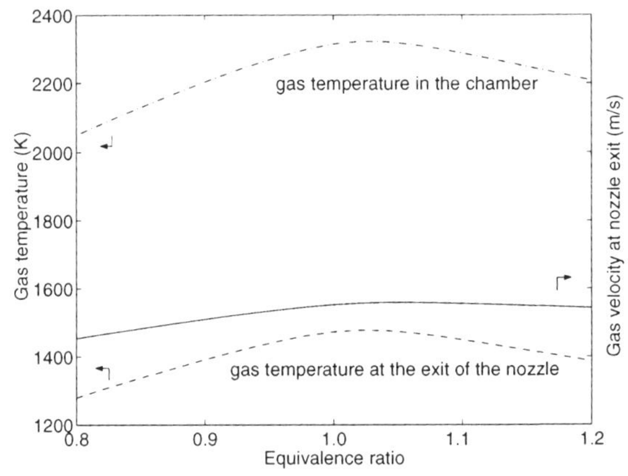

Figure 8.2. Gas properties in the combustion chamber (dash-dot line: gas temperature) and at the exit of the nozzle (dashed line: gas temperature, solid line: gas velocity) - $P = 9$ *bar* and $\varphi = 0.8 \sim 1.2$.

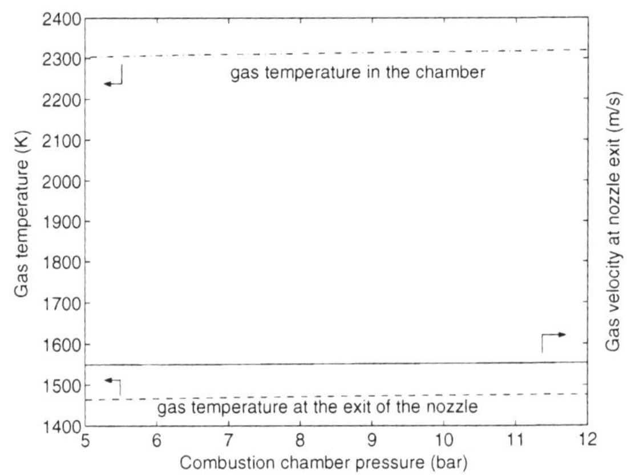

Figure 8.3. Gas properties in the combustion chamber (dash-dot line: gas temperature) and at the exit of the nozzle (dashed line: gas temperature, solid line: gas velocity) - $\varphi = 1.0$ and $P = 5 \sim 12$ *bar*.

behavior is also observed for the temperature profiles of the particles (Figure 8.5).

While we concluded from Figure 8.3 that chamber pressure does not affect gas velocity, the strong effect of chamber pressure on particle velocity is not unexpected. This is because chamber pressure influences gas density, and thus, particle velocity (Eq.8.15). Figure 8.6 shows gas density at the exit of the nozzle as a function of the chamber pressure. The gas density is nearly proportional to the chamber pressure.

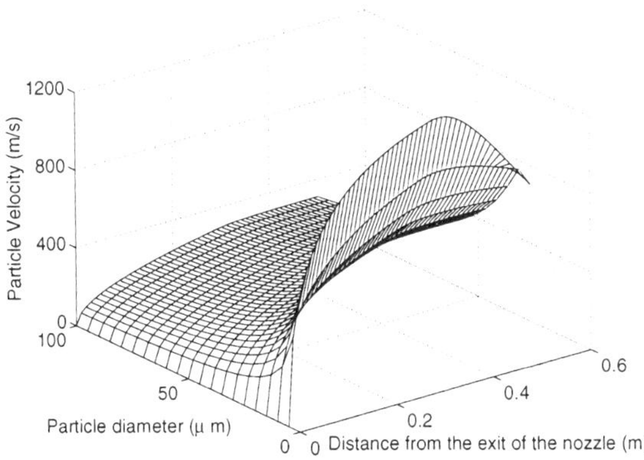

Figure 8.4. Axial mean velocity profile of particles from the exit of the nozzle for different particle sizes - $P = 9\ bar$ and $\varphi = 1.0$.

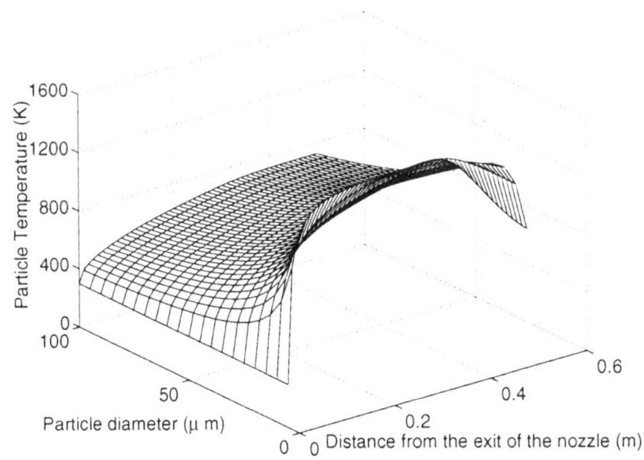

Figure 8.5. Axial mean temperature profile of particles from the exit of the nozzle for different particle sizes - $P = 9\ bar$ and $\varphi = 1.0$.

Figures 8.7 and 8.8 show the particle velocity and temperature at the 0.3 m standoff versus particle size. As particle diameter increases, both particle velocity and temperature increase first, and then decrease gradually. This phenomenon is caused by the low acceleration force for fine particles in the barrel and large deceleration force for coarse particles in the supersonic free jet. The fact that the maximum values of both velocity and temperature correspond to the median particle size implies that particle size of the feedstock is one of the key parameters deciding coating quality and motivate controlling average particle velocity and temperature.

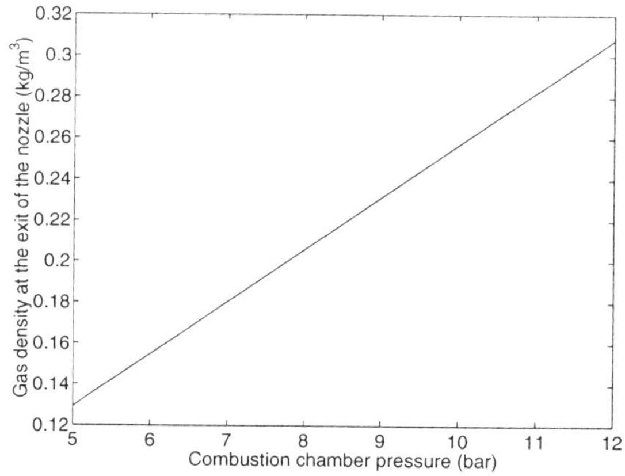

Figure 8.6. Gas density at the exit of the nozzle - $\varphi = 1.0$ and $P = 5 \sim 12\ bar$.

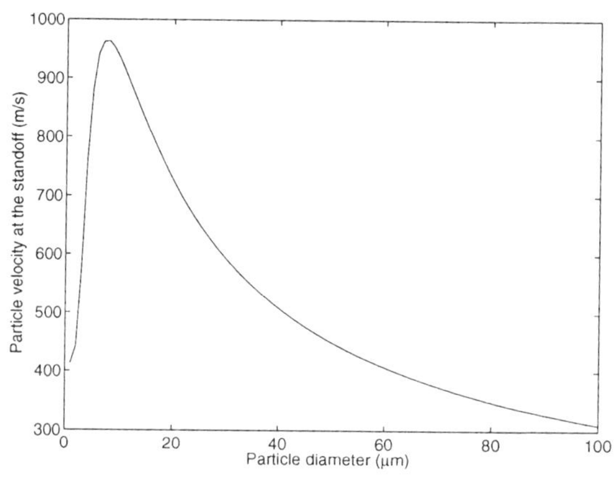

Figure 8.7. Axial mean particle velocity at the point of impact on substrate as a function of particle size under operating condition $P = 9\ bar$ and $\varphi = 1.0$.

Figures 8.9 and 8.10 show the influence of chamber pressure and equivalence ratio on velocity and temperature profiles of a single particle with $d_p = 40\ \mu m$. In Figure 8.9, we fix the equivalence ratio and vary the chamber pressure. As the chamber pressure increases, both particle velocity and particle temperature increase. However, particle velocity is more sensitive to chamber pressure than particle temperature. In Figure 8.10, we change the equivalence ratio with constant chamber pressure. While the particle temperature decreases as the equivalence ratio departs from 1, the particle velocity changes very little. From these results it can be concluded that the particle velocity is primarily influenced by the combustion chamber pressure and the particle temperature depends

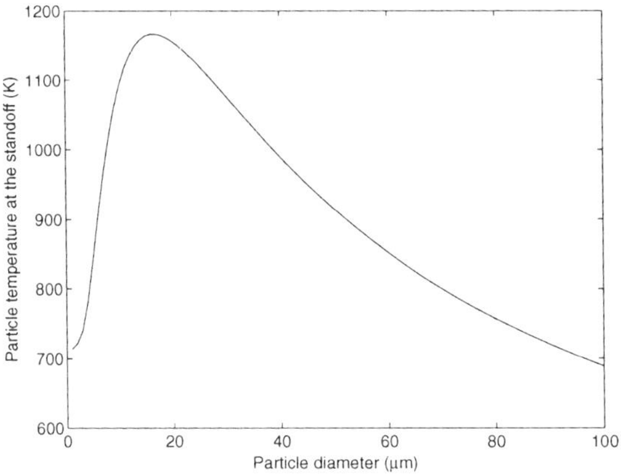

Figure 8.8. Axial mean particle temperature at the point of impact on substrate as a function of particle size under operating condition $P = 9\ bar$ and $\varphi = 1.0$.

on the equivalence ratio, which is in good agreement with available experimental data (Swank et al., 1994a; Swank et al., 1994b). As we mentioned before, although chamber pressure has a minimal influence on gas velocity, it does influence particle velocity because, the drag force on particle depends strongly on gas density, which, in turn, depends on chamber pressure. Note that combustion chamber pressure is dependent on the total flow rate of air and fuel, so we can use the total flow rate and the relative flow rate of air and fuel to control the particle velocity and particle temperature at the point of impact on substrate, respectively.

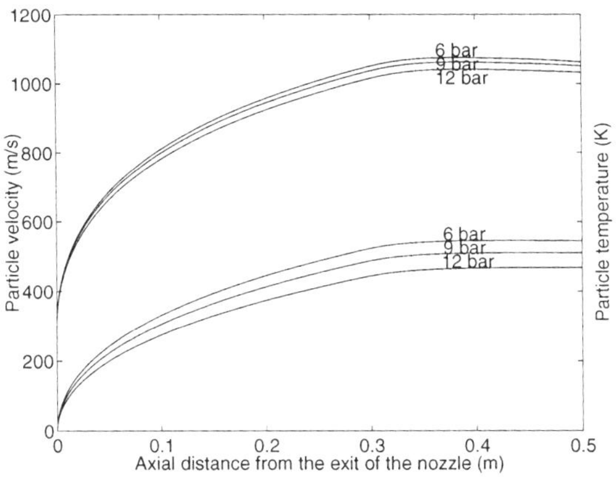

Figure 8.9. Velocity and temperature profiles for particle with size $d_p = 40\ \mu m$ under operating conditions $\varphi = 1.0$ and different chamber pressures $P = 6$, 9 and 12 bar.

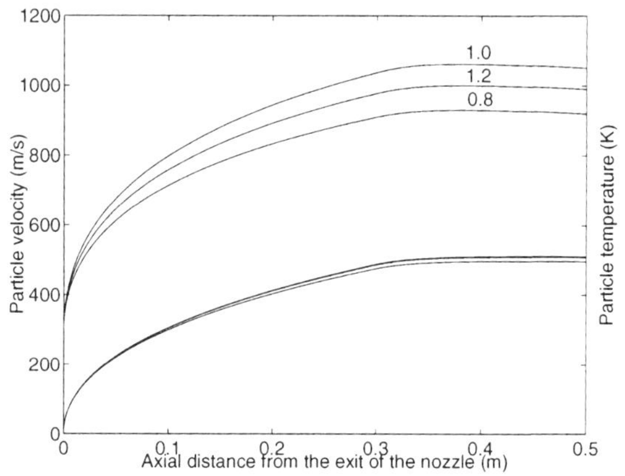

Figure 8.10. Velocity and temperature profiles for particle with size $d_p = 40\ \mu m$ under operating conditions $P = 9\ bar$ and different equivalence ratios $\varphi = 0.8$, 1.0 and 1.2.

Table 8.3 shows the volume based average velocities and temperatures for particles with $d_{10} = 15\ \mu m$, $d_{10} = 35\ \mu m$ and $d_{10} = 77\ \mu m$ at the point of impact on substrate under five different operating conditions. The same conclusion as above can be drawn for the relationship of combustion chamber pressure and fuel/oxy ratio and particle velocity and particle temperature.

P (bar)	φ	$\bar{v}_f$ (m/s)	T_f (K)
6.0	1.0	532.0	1045.0
9.0	0.8	551.2	918.8
9.0	1.0	571.6	1056.5
9.0	1.2	568.7	993.0
12.0	1.0	603.3	1062.2

Table 8.3. Volume based average particle velocity and temperature at the point of impact on substrate under five different operating conditions.

4. Feedback Control - Closed-loop Simulations

Based on the model predictions, and the available experimental results, the control problem for the HVOF system is formulated as the one of regulating the temperature and velocity of the particles at impact on the substrate (these are the variables that directly influence coating microstructure and porosity which determine coating strength and hardness) by manipulating the combustion chamber pressure (total inlet flow of combustion gases) and the oxygen/fuel ratio. Our open-loop simulation results show that ρ_g, which strongly affects v_p and T_g, which strongly affects T_p, can be almost independently adjusted by manipulating the combustion chamber pressure, which is related to total gas flow rate, $u_1(t)$, and the equivalence ratio, $u_2(t)$, respectively.

Owing to the almost decoupled nature of the manipulated input and controlled output pairs, two proportional integral (PI) controllers are used to regulate the process. Specially, the controls have the following form:

$$\begin{aligned} \dot{\zeta}_i &= y_{sp_i} - y_i, \;\; \zeta_i(0) = 0, \;\; i = 1, 2 \\ u_i &= K_{c_i}(y_{sp_i} - y_i) + \frac{1}{\tau_c}\zeta_i + u_{0_i} \end{aligned} \tag{8.28}$$

where y_{sp_i} is the specified output, y_1 is the averaged particle velocity and y_2 is the averaged particle temperature (both computed at the point of impact on substrate using Eq. 8.27), K_{c_i} is the proportional gain and τ_{c_i} is the integral time constant.

In an experimental application, an ensemble average of particle velocity can be measured by optics-based methods, such as Laser Doppler Velocimeter (LDV) and Laser-two-focus (L2F), which have no interference with the gas/particle field (Wagner et al., 1984; Fincke et al., 1990; Knight et al., 1994). Particle temperature is determined by measuring the thermal radiation intensity at two different wavelengths (also known as two color pyrometry) (Fincke et al., 2001a). In the closed loop simulations, $y_1(t)$ and $y_2(t)$ are computed from appropriate averaging of the individual particle velocity and temperature data obtained from the process model. To account for powder size distribution, 100 particles with different size are traced simultaneously to obtain the average velocity and temperature. This requires solving simultaneously 300 ordinary differential equations. The parameters used in the closed-loop simulations are shown in table 8.4; the controller parameter were chosen to achieve smooth controlled output transient responses without using unrealistically large control action.

Figures 8.11 and 8.12 show the evolution of the controlled and manipulated variables, respectively, in the case of requesting a 10% change in the set-point values. The particles achieve the specified average velocity and average temperature in about 20 seconds, which validates the feasibility of real-time implementation of a feedback controller on the HVOF system.

To show that the proposed formulation of the control problem (which explicitly accounts for the effect of powder size distribution) leads to a solution of the control problem that is superior (with respect to the

K_{c_1}	1000
K_{c_2}	0.001
τ_1	0.2
τ_2	0.2
d_{10}	27 μm
d_{50}	80 μm
d_{90}	144 μm
ϕ	1.0

Table 8.4. Controller parameters used in the simulation.

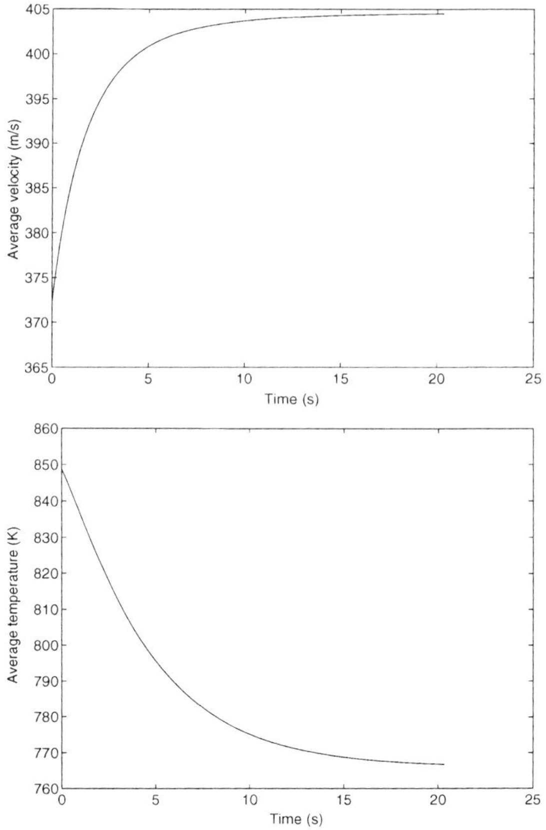

Figure 8.11. Profiles of weight-based average particle velocity (top plot) and particle temperature (bottom plot) with respect to time - Control problem formulation accounting for powder size distribution.

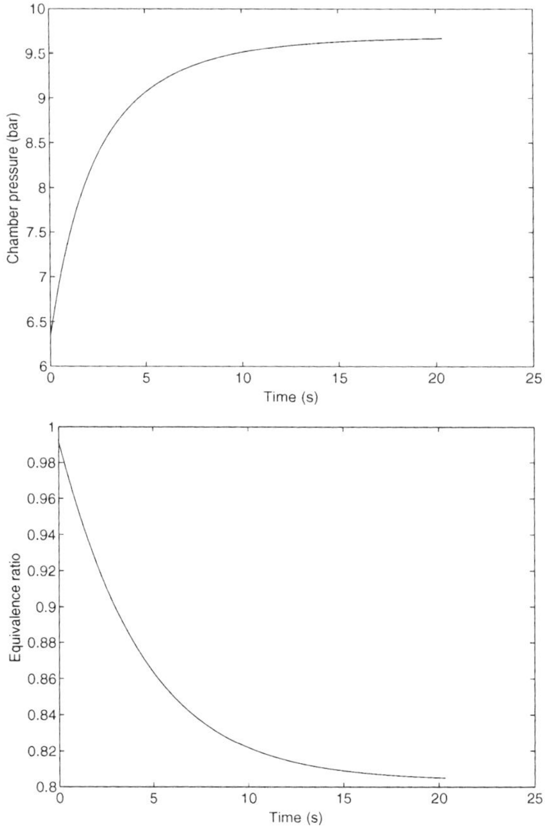

Figure 8.12. Profiles of chamber pressure (top plot) and equivalence ratio (bottom plot) with respect to time - Control problem formulation accounting for powder size distribution.

control action needed to achieve the desired control objectives) to a solution that assumes a monodisperse powder size distribution, we also implemented on the process model the two PI controllers under the assumption of monodisperse powder size distribution (the single particle size was taken to be $d_p = 20 \mu m$). The corresponding controlled and manipulated variables are given in figures 8.13 and 8.14, respectively. The results show that the average particle velocity increases by 22.85% and the average particle temperature decreases by 6.95%. Therefore, the desired objectives of 10% change in the set-point values are not achieved, which makes clear the need to account for the effect of powder size distribution in the control problem formulation and solution.

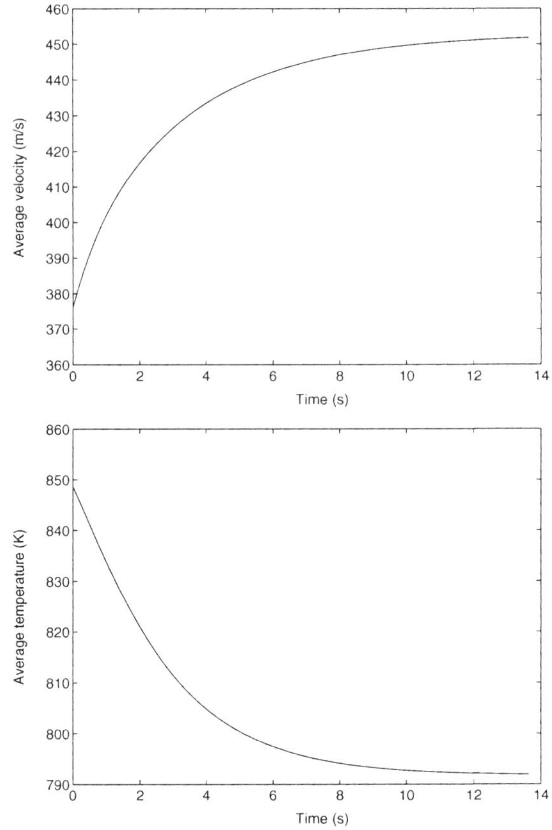

Figure 8.13. Profiles of weight-based average particle velocity (top plot) and particle temperature (bottom plot) with respect to time - Control problem formulation without accounting for powder size distribution.

Finally, to test the robustness of the proposed control problem formulation and of the controller, we considered controlling the HVOF process in the presence of disturbances in the spray distance and in the initial velocity of the injected particles.

Figures 8.15 and 8.16 show the controlled output and manipulated input profiles in the presence of 30% disturbance in the spray distance which occurs at t = 10 *s*. Without control, the process jumps to a new steady state in a very short time (owing to the very short time of particle fight), and both the particle velocity and particle temperature drop instantaneously. With feedback control, the process outputs (dashed lines) move gradually and finally achieve their original steady state values in about 30 seconds.

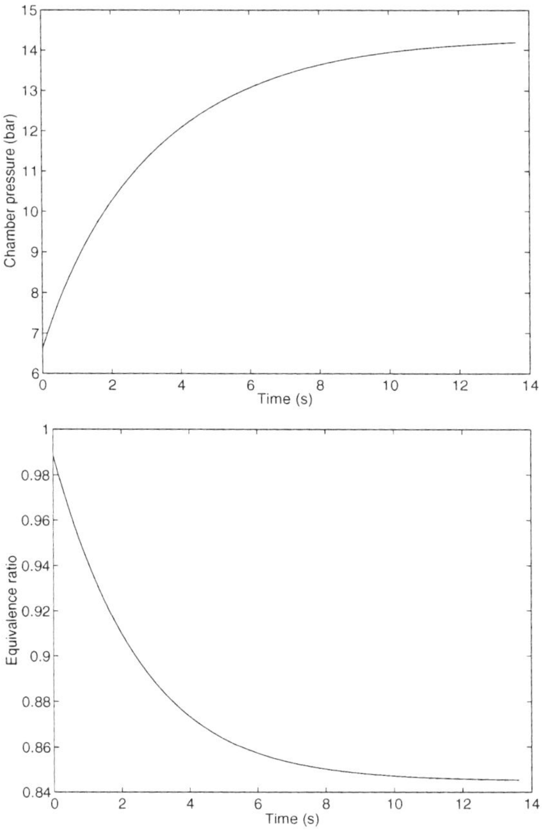

Figure 8.14. Profiles of chamber pressure (top plot) and equivalence ratio (bottom plot) with respect to time - Control problem formulation without accounting for powder size distribution.

Figures 8.17 and 8.18 show the controlled output and manipulated input profiles in the presence of 20 m/s change in initial particle velocity at t = 10 s. Without control, the system jumps to a new steady state in a very short time; this is due to the very short time of particle flight. The particle velocity slightly increases and the particle temperature decreases by about 15 K. The drop of particle temperature is explained by the shortened residence time of particles in the gas, which is caused by the increased time-averaged particle velocity. Such a temperature change can have a significant effect on the molten state of the particle and the resulting coating microstructure. Under feedback control, the process outputs reach their original steady state values in about 35 seconds.

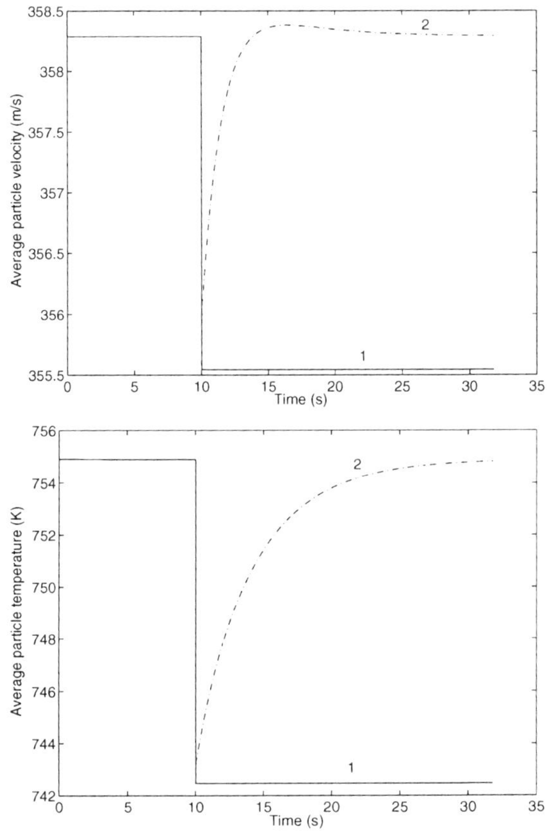

Figure 8.15. Profiles of weight-based average particle velocity (top plot) and particle temperature (bottom plot) with respect to time in the presence of a 30% disturbance on spray distance. Open-loop system (solid line) and closed-loop system (dashed line).

In summary, our simulation results show that the proposed formulation of the control problem, which explicitly accounts for the effect of powder size distribution, leads to effective control of particle velocity and temperature in the HVOF process. The proposed approach was also shown to be robust with respect to significant variations in the operating conditions.

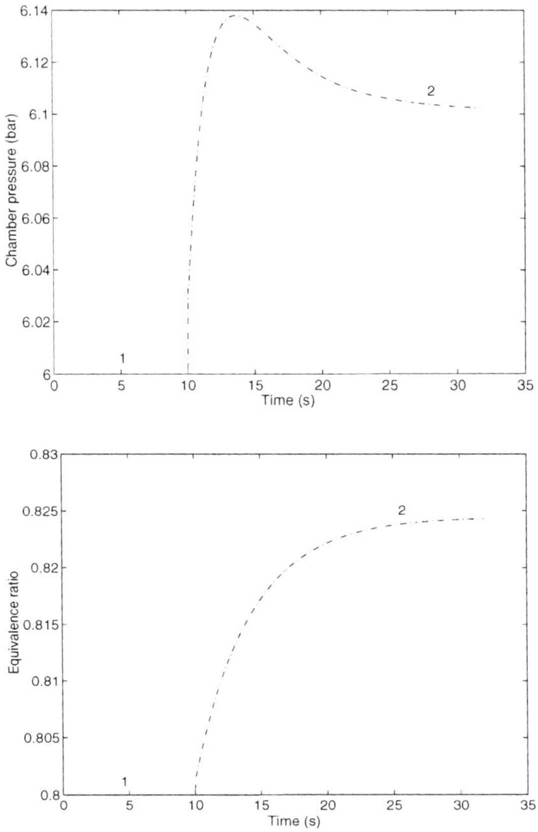

Figure 8.16. Profiles of chamber pressure (top plot) and equivalence ratio (bottom plot) with respect to time in the presence of a 30% disturbance on spray distance. Open-loop system (solid line) and closed-loop system (dashed line).

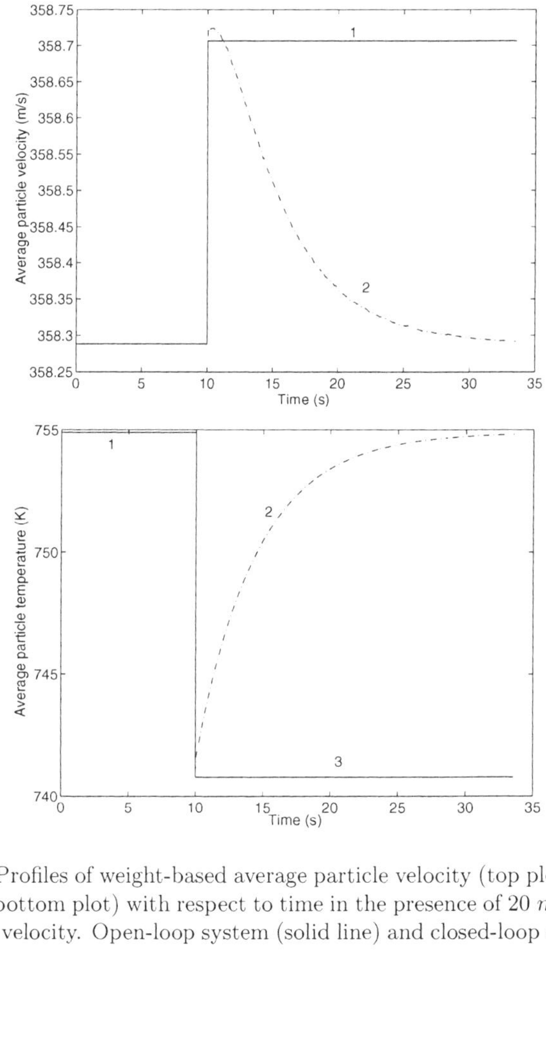

Figure 8.17. Profiles of weight-based average particle velocity (top plot) and particle temperature (bottom plot) with respect to time in the presence of 20 m/s variation in initial particle velocity. Open-loop system (solid line) and closed-loop system (dashed line).

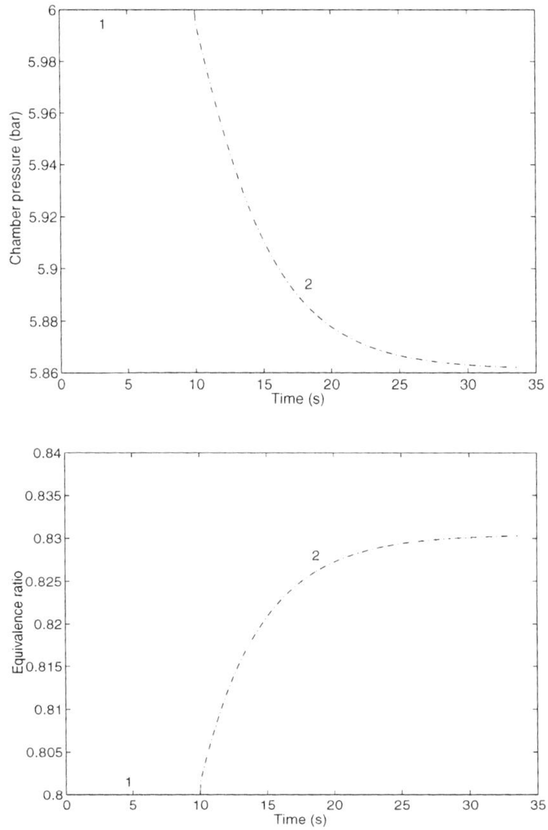

Figure 8.18. Profiles of chamber pressure (top plot) and equivalence ratio (bottom plot) with respect to time in the presence of 20 m/s variation in initial particle velocity. Open-loop system (solid line) and closed-loop system (dashed line).

References

Akhtar, M. K., Xiong, Y., and Pratsinis, S. E. (1991). Vapor synthesis of titania powder by titanium tetrachloride oxidation. *AIChE J.*, 37:1561–1570.

Alvarez, J., Alvarez, J. J., and Suraez, R. (1991). Nonlinear bounded control for a class of continuous agitated tank reactors. *Chem. Eng. Sci.*, 46:3235–3249.

Annavarapu, S., Apelian, D., and Lawley, A. (1990). Spray casting of steel strip: Process analysis. *Metall. Trans. A.*, 21:3237–3256.

Antoniades, C. and Christofides, P. D. (1999). Feedback control of nonlinear differential difference equation systems. *Chem. Eng. Sci.*, 54:5677–5709.

Balas, M. J. (1995). Finite-dimensional direct adaptive control for discrete-time infinite-dimensional linear systems. *J. Math. Anal. Appl.*, 196:153–171.

Bird, R. B., W., S. E., and Lightfoot, E. N. (1960). *Transport Phenomena.* John Wiley & Sons, New York.

Bohren, C. F. and Huffman, D. R. (1983). *Absorption and Scattering of Light by Small Particles.* Wiley, New York.

Brock, J. R., Kuhn, P. J., and Zehavi, D. (1986). Condensation aerosol formation and growth in a laminar coaxial jet: Experimental. *J. Aerosol Sci.*, 17:11–22.

Byrnes, C. I. (1987). Adaptive stabilization of infinite dimensional linear systems. In *Proceedings of 26th IEEE Conference on Decision and Control*, pages 1435–1440, Los Angeles, CA.

Cheng, D., Trapaga, G., McKelliget, J. W., and Lavernia, E. J. (2001a). Mathematical modeling of high velocity oxygen fuel thermal spraying: an overview. *Key Engineering Materials.*

Cheng, D., Xu, Q., Trapaga, G., and Lavernia, E. J. (2001b). The effect of particle size and morphology on the in-flight behavior of particles during high-velocity oxyfuel thermal spraying. *Metallurgical and Materials Transactions B*, 32:525–535.

Chiu, T. and Christofides, P. D. (1999a). Nonlinear control of particulate processes. *AIChE J.*, 45:1279–1297.

Chiu, T. and Christofides, P. D. (1999b). Robust nonlinear control of a continuous crystallizer. *Comp. & Chem. Eng.*, 23(s):249–252.

Chiu, T. and Christofides, P. D. (2000). Robust control of particulate processes using uncertain population balances. *AIChE J.*, 46:266–280.

Christofides, P. D. (1998). Robust control of parabolic PDE systems. *Chem. Eng. Sci.*, 53:2949–2965.

Christofides, P. D. (2001). *Nonlinear and Robust Control of PDE Systems: Methods and Applications to Transport-Reaction Processes.* Birkhäuser, Boston.

Christofides, P. D. and Chiu, T. (1997). Nonlinear control of particulate processes. In *AIChE Annual Meeting, paper 196a, Los Angeles, CA.*

Christofides, P. D. and Daoutidis, P. (1996). Feedback control of hyperbolic PDE systems. *AIChE J.*, 42:3063–3086.

Christofides, P. D. and Daoutidis, P. (1997). Finite-dimensional control of parabolic PDE systems using approximate inertial manifolds. *J. Math. Anal. Appl.*, 216:398–420.

Christofides, P. D. and Daoutidis, P. (1998). Robust control of hyperbolic PDE systems. *Chem. Eng. Sci.*, 53:85–105.

Christofides, P. D. and Teel, A. R. (1996). Singular perturbations and input-to-state stability. *IEEE Trans. Autom. Contr.*, 41:1645–1650.

Colonius, F. and Kliemann, W. (1993). Some aspects of control of systems as dynamical systems. *J. Dynamics and Differential Equations*, 5:469–479.

Crow, E. L. and Shimizu, K. (1988). *Lognormal Distributions: Theory and Applications.* Marcel Dekker, New York, USA.

Crowe, C. T. and Stock, D. E. (1976). A computer solution for two-dimensional fluid-particle flows. *International Journal for Numerical Methods in Engineering*, 10:185–196.

Daoutidis, P. and Henson, M. (2001). Dynamics and control of cell populations. In *Proceedings of 6th International Conference on Chemical Process Control*, pages 308–325, Tucson, AZ.

Dimitratos, J., Elicabe, G., and Georgakis, C. (1994). Control of emulsion polymerization reactors. *AIChE J.*, 40:1993–2021.

Eaton, J. W. and Rawlings, J. B. (1990). Feedback control of chemical processes using on-line optimization techniques. *Comp. Chem. Eng.*, 14:469–479.

El-Farra, N., Armaou, A., and Christofides, P. D. (2002). Analysis and control of parabolic PDE systems with input constraints. *Automatica,* to appear.

El-Farra, N. H., Chiu, T., and Christofides, P. D. (2001). Analysis and control of particulate processes with input constraints. *AIChE J.*, 47:1849–1865.

El-Farra, N. H. and Christofides, P. D. (2001a). Integrating robustness, optimality, and constraints in control of nonlinear processes. *Chem. Eng. Sci.*, 56:1–28.

El-Farra, N. H. and Christofides, P. D. (2001b). Robust near-optimal output feedback control of nonlinear systems. *Intr. J. Contr.*, 74:133–157.

Fincke, J. R., Haggard, D. C., and Swank, W. D. (2001a). Particle temperature measurement in the thermal spray process. *Journal of Thermal Spray Technology*, 10:255–266.

Fincke, J. R., Swank, W. D., Bewley, R. L., Haggard, D. C., Gevelber, M., and Wroblewski, D. (2001b). Diagnostics and control in the thermal spray process. *Surface and Coatings Technology*, 146-147:537–543.

Fincke, J. R., Swank, W. D., and Jeffrey, C. L. (1990). Simultaneous measurement of particle size, velocity and temperature in thermal plasmas. *IEEE Transactions on Plasma Science*, 18:948–957.

Fisher, W. R., Doherty, M. F., and Douglas, J. M. (1985). Steady-state control as a prelude to dynamic control. *Chem. Eng. Res. Des.*, 6:353–357.

Foias, C. and Témam, R. (1988). Algebraic approximation of attractors: The finite dimensional case. *Physica D*, 32:163–182.

Freeman, R. A. and Kokotovic, P. V. (1996). *Robust Nonlinear Control Design: State-Space and Lyapunov Techniques.* Birkhauser, Boston.

Friedlander, S. K. (1977). *Smoke, Dust, and Haze: Fundamentals of Aerosol Behavior.* Wiley, New York.

Friedlander, S. K. (1983). Dynamics of aerosol formation by chemical reaction. *Annals. N.Y. Acad. Sci.*, 83:354–364.

Fukunaga, K. (1990). *Introduction to statistical pattern recognition.* Academic Press, New York.

Ganser, G. H. (1993). A rational approach to drag prediction of spherical and nonspherical particles. *Powder Technology*, 77:143–152.

Geankoplis, C. J. (1993). *Transport Processes and Unit Operations (3rd ed.).* Prentice-Hall, New Jersey, USA.

Gelbard, F. and Seinfeld, J. H. (1978). Numerical solution of the dynamic equation for particulate processes. *J. Comp. Phys.*, 28:357–375.

Gelbard, F., Tambour, Y., and Seinfield, J. H. (1980). Sectional representation of simulating aerosol dynamics. *J. Coll. Inter. Sci.*, 68:363–382.

Gordon, S. and McBride, B. J. (1994). *Computer Program for Calculation of Complex Chemical Equilibrium Compositions and Applications.* NASA Reference Publication 1311, Lewis Researach Center, Cleveland, Ohio, USA.

Gourlaouen, V., Verna, E., and Beaubien, P. (2000). Influence of flame parameters on stainless steel coatings properties. In *Thermal Spray: Surface Engineering Via Applied Research, Proceedings of the International Thermal Spray Conference*, pages 487–493, Montreal, QC, Canada.

Hassan, B., Lopez, A. R., and Oberkampf, W. L. (1998). Computational analysis of a three-dimensional high-velocity oxygen fuel (HVOF) thermal spray torch. *Journal of Thermal Spray Technology*, 7:71–77.

Hill, P. J. and Ng, K. M. (1996). New discretization procedure for the agglomeration equation. *AIChE J.*, 42:727–741.

Holmes, P., Lumley, J. L., and Berkooz, G. (1996). *Turbulence, Coherent Structures, Dynamical Systems and Symmetry.* Cambridge University Press, New York.

Hounslow, M. J. (1990). A discretized population balance for continuous systems at steady-state. *AIChE J.*, 36:106–116.

Hulburt, H. M. and Katz, S. (1964). Some problems in particle technology: A statistical mechanical formulation. *Chem. Eng. Sci.*, 19:555–574.

Isidori, A. (1995). *Nonlinear Control Systems: An Introduction.* Springer-Verlag, Berlin-Heidelberg, third edition.

Jerauld, G. R., Vasatis, Y., and Doherty, M. F. (1983). Simple conditions for the appearance of sustained oscillations in continuous crystallizers. *Chem. Eng. Sci.*, 38:1675–1681.

Jones, A. G. (1974). Optimal operation of a batch cooling crystallizer. *Chem. Eng. Sci.*, 29:1075–1087.

Jones, A. G. and Mullin, J. W. (1974). Programmed cooling crystallization of potassium sulphate solutions. *Chem. Eng. Sci.*, 29:105–118.

Kalani, A. and Christofides, P. D. (1999). Nonlinear control of spatially-inhomogeneous aerosol processes. *Chem. Eng. Sci.*, 54:2669–2678.

Kalani, A. and Christofides, P. D. (2000). Modeling and control of a titania aerosol reactor. *Aer. Sci. & Tech.*, 32:369–391.

Kalani, A. and Christofides, P. D. (2002). Simulation, estimation and control of size distribution in aerosol processes with simultaneous reaction, nucleation, condensation and coagulation. *Comp. & Chem. Eng.*, 26:1153–1169.

Kapoor, N. and Daoutidis, P. (1999). On the dynamics of nonlinear systems with input constraints. *Chaos*, 9:88–94.

Kapoor, N., Teel, A. R., and Daoutidis, P. (1998). An anti-windup design for linear systems with input saturation. *Automatica*, 34:559–574.

Kelley, K. (1960). *High-Temperature Heat-Content, Heat-Capacity and Entropy Data for the Elements and Inorganic Compounds.* United States Government Printing Office, Washington.

Khalil, H. K. (1996). *Nonlinear Systems.* Macmillan Publishing Company, New York, second edition.

Knight, R., Smith, R. W., Xiao, Z., and Hoffman, T. T. (1994). Particle velocity measurements in HVOF and APS systems. In *Thermal Spray Industrial Applications Conference Proceedings, ASM international*, pages 331–336, Materials Park, Ohio.

Kobata, A., Kusakabe, K., and Morooka, S. (1991). Growth and transformation of TiO_2 crystallites in aerosol reactor. *AIChE J.*, 37:347–359.

Kokotovic, P. V., Khalil, H. K., and O'Reilly, J. (1986). *Singular Perturbations in Control: Analysis and Design.* Academic Press, London.

Kothare, M. V., Campo, P. J., Morari, M., and Nett, C. N. (1994). A unified framework for the study of anti-windup designs. *Automatica*, 30:1869–1883.

Kumar, A., Christofides, P. D., and Daoutidis, P. (1998). Singular perturbation modeling and control of nonlinear two-time-scale processes with non-explicit time-scale multiplicity. *Chem. Eng. Sci.*, 53:1491–1504.

Kumar, S. and Ramkrishna, D. (1996a). On the solution of population balance equations by discretization-I. a fixed pivot technique. *Chem. Eng. Sci.*, 51:1311–1332.

Kumar, S. and Ramkrishna, D. (1996b). On the solution of population balance equations by discretization-II. a moving pivot technique. *Chem. Eng. Sci.*, 51:1333–1342.

Kurtz, M. J., Zhu, G.-Y., Zamamiri, A., Henson, M. A., and Hjortso, M. A. (1998). Control of oscillating microbial cultures described by population balance models. *Ind. Eng. Chem. Res.*, 37:4059–4070.

Landgrebe, J. D. and Pratsinis, S. E. (1990). A discrete sectional model for particulate production by gas phase chemical reaction and aerosol coagulation in the free molecular regime. *J. Coll. Inter. Sci.*, 139:63–86.

Lau, M. L., Gupta, V. V., and Lavernia, E. J. (1998a). Mathematical modeling of particle behavior of nanocrystalline Ni during high velocity oxy-fuel thermal spray. *Nanostructured Materials*, 10:715–722.

Lau, M. L., Gupta, V. V., and Lavernia, E. J. (1999). Particle behavior of nanocrystalline 316-stainless steel during high velocity oxy-fuel thermal spray. *Nanostructured Materials*, 12:319–322.

Lau, M. L., Jiang, H. G., Nuchter, W., and Lavernia, E. J. (1998b). Thermal spraying of nanocrystalline Ni coatings. *Physica Status Solidi A - Applied Research*, 166:257–268.

Lee, K. W. (1984). Log-normally preserving size distribution for brownian coagulation in the free-molecule regime. *Aerosol Sci. Tech.*, 3:53–62.

Lei, S. J., Shinnar, R., and Katz, S. (1971). The stability and dynamic behavior of a continuous crystallizer with a fines trap. *AIChE J.*, 17:1459–1470.

Li, M. and Christofides, P. D. (2002). Feedback control of nanostructured coatings processing accounting for powder size distribution. In *American Control Conference*, pages 1019–1027; To appear in the Journal of Thermal Spray Technology.

Lin, Y. and Sontag, E. D. (1991). A universal formula for stabilization with bounded controls. *Systems & Control Letters*, 16:393–397.

Marthur, P., Annavarapu, S., Apelian, D., and Lawley, A. (1991). Spray casting: An integral model for process understanding and control. *Mat. Sci. Eng. A*, 142:261–276.

Miller, S. M. and Rawlings, J. B. (1994). Model identification and control strategies for batch cooling crystallizers. *AIChE J.*, 40:1312–1327.

Moreau, C. and Leblanc, L. (2001). Optimization and process control for high performance thermal spray coatings. *Key Engineering Materials*, 197:27–57.

Pawlowski, L. (1995). *The Science and Engineering of Thermal Spraying Coatings.* John Wiley & Sons, Chichester, England.

Power, G. D., Barber, T. J., and Chiappetta, L. M. (1992). Analysis of a high velocity oxygen-fuel(HVOF) thermal torch. In *the 28th AIAA/SAE/ASME/ASEE Joint Propulsion Conference and Exhibit*, pages AIAA92–3598, Nashville, Tennessee.

Pratsinis, S. E. (1988). Simultaneous nucleation, condensation, and coagulation in aerosol reactors. *J. Coll. Inter. Sci.*, 124:416–426.

Pratsinis, S. E. (1989). Particle production by gas-to-particle conversion in turbulent flows. *J. Aerosol Sci.*, 20:1461–1464.

Pratsinis, S. E., Bai, H., Biswas, P., Frenklach, M., and Mastrangelo, S. V. R. (1990). Kinetics of titanium IV chloride oxidation. *J. Am. Ceram. Soc.*, 73:2158–9999.

Ramabhadran, T. E. and Seinfield, J. H. (1975). Self-preserving theory of particulate systems. *Chem. Eng. Sci.*, 30:1019–1025.

Ramkrishna, D. (1985). The status of population balances. *Rev. Chem. Eng.*, 3:49–95.

Randolph, A. D., Chen, L., and Tavana, A. (1987). Feedback control of CSD in a KCl crystallizer with a fines dissolver. *AIChE J.*, 33:583–591.

Randolph, A. D. and Larson, M. A. (1988). *Theory of Particulate Processes.* Academic press, Second edition, San Diego.

Rao, C. V. and Rawlings, J. B. (1999). Steady states and constraints in model predictive control. *AIChE J.*, 45:1266–1278.

Rawlings, J. B., Miller, S. M., and Witkowski, W. R. (1993). Model identification and control of solution crystallization processes. *I & EC Res.*, 32:1275–1296.

Rawlings, J. B. and Ray, W. H. (1987a). Emulsion polymerization reactor stability: simplified model analysis. *AIChE J.*, 33:1663–1667.

Rawlings, J. B. and Ray, W. H. (1987b). Stability of continuous emulsion polymerization reactors: a detailed model analysis. *Chem. Eng. Sci.*, 42:2767–2777.

Roberson, J. A. and Crowe, C. T. (1997). *Engineering Fluid Dynamics (6th ed.).* John Wiley & Sons, New York, USA.

Rohani, S. and Bourne, J. R. (1990). Self-tuning control of crystal size distribution in a cooling batch crystallizer. *Chem. Eng. Sci.*, 12:3457–3466.

Scokaert, P. O. M. and Rawlings, J. B. (1999). Feasibility issues in linear model predictive control. *AIChE J.*, 45:1649–1259.

Semino, D. and Ray, W. H. (1995a). Control of systems described by population balance equations-I. controllability analysis. *Chem. Eng. Sci.*, 50:1805–1824.

Semino, D. and Ray, W. H. (1995b). Control of systems described by population balance equations-II. emulsion polymerization with constrained control action. *Chem. Eng. Sci.*, 50:1825–1839.

Sepulchre, R., Jankovic, M., and Kokotovic, P. (1997). *Constructive Nonlinear Control.* Springer-Verlag, Berlin-Heidelberg.

Shvartsman, S. Y. and Kevrekidis, I. G. (1998). Nonlinear model reduction for control of distributed parameter systems: A computer assisted study. *AIChE J.*, 44:1579–1595.

Stephanopoulos, G. (1983). Synthesis of control systems for chemical plants-a challenge for creativity. *Comp. & Chem. Eng.*, 7:331–365.

Swank, W. D., Fincke, J. R., Haggard, D. C., and Irons, G. (1994a). HVOF gas low field characteristics. In *Proceedings of the 7th National Thermal Spray Conference*, pages 313–318, Boston, Massachusetts.

Swank, W. D., Fincke, J. R., Haggard, D. C., Irons, G., and Bullock, R. (1994b). HVOF particle flow field characteristics. In *Proceedings of the 7th National Thermal Spray Conference*, pages 319–324, Boston, Massachusetts.

Tawfik, H. H. and Zimmerman, F. (1997). Mathematical modeling of the gas and powder flow in HVOF systems. *Journal of Thermal Spray Technology*, 6:345–352.

Tellkamp, M. L., Lau, A., and Lavernia, E. J. (1997). Thermal spraying of nanocrystalline inconel 718. *Nanostructured Materials*, 9:489–492.

Temam, R. (1988). *Infinite-Dimensional Dynamical Systems in Mechanics and Physics*. Springer-Verlag, New York.

Wagner, N., Gnadig, K., Kreye, H., and Kronewetter, H. (1984). Particle velocity in hypersonic flame spraying of $WC - Co$. *Surface Technology*, 22:61–71.

Wang, G.-X., Prasad, V., and Sampath, S. (2001). Rapid solidification in thermal spray deposition: Microstructure and modelling. *Sadhana-Academy Proceedings in Engineering Sciences*, 26:35–57.

Wen, J. T. and Balas, M. J. (1989). Robust adaptive control in hilbert space. *J. Math. Anal. Appl.*, 143:1-26.

William, M. M. R. and Loyalka, S. K. (1991). *Aerosol Science: Theory & Practice*. Pergamon Press, Oxford, England.

Xiong, Y. and Pratsinis, S. E. (1991). Gas phase production of particles in reactive turbulent flows. *J. Aerosol Sci.*, 12:637–655.

Xu, Q. and Lavernia, E. J. (2001). Influence of nucleation and growth phenomena on microstructural evolution during droplet-based deposition. *Acta Materialia*, 49:18.

Yang, X. and Eidelman, S. (1996). Numerical analysis of a high-velocity oxygen-fuel thermal spray system. *Journal of Thermal Spray Technology*, 5:175–184.

Ydstie, E. B. and Alonso, A. A. (1997). Process systems and passivity via the Clausius-Planck inequality. *Syst. & Contr. Lett.*, 30:253–264.

Zagorski, A. V. and Stadelmaier, F. (2001). Full-scale modelling of a thermal spray process. *Surface and Coatings Technology*, 146-147:162–167.

APPENDIX 9.A: Proofs of Chapter 2

Proof of Proposition 2.1: To simplify the notation of the proof, we assume that the growth rate $G(x,r)$ is independent of r. Furthermore, since the nonlinear terms $w(n,x,r), a(n,r,x)$ are smooth functions of their arguments, they can be exactly expanded in an infinite series as follows:

$$\begin{aligned} w(n,x,r) &= \sum_{k=1}^{\infty} b_k(x,a_k,t)\phi_k(r) \\ a(n,r,x) &= \sum_{k=1}^{\infty} c_k(x,a_k,t)\phi_k(r) \end{aligned} \tag{9.A.1}$$

where $b_k(x,a_k,t), c_k(x,a_k,t)$ are coefficients and $\phi_k(r)$, $k=1,\ldots,\infty$ is the complete set of basis functions used to expand $n(r,t)$ in Eq.2.17. Using that $G(x,r)$ is independent of r and the expansions of Eq.9.A.1, the infinite dimensional system of Eq.2.19 can be written as:

$$\begin{aligned} \int_0^{r_{max}} \psi_\nu(r)\sum_{k=1}^{\infty}\phi_k(r)\frac{\partial a_k(t)}{\partial t}dr &= -G(x)\sum_{k=1}^{\infty}a_k(t)\int_0^{r_{max}}\psi_\nu(r)\frac{\partial \phi_k(r)}{\partial r}dr \\ &\quad +\int_0^{r_{max}}\psi_\nu(r)\sum_{k=1}^{\infty}b_k(x,a_k,t)\phi_k(r)dr, \\ &\quad \nu=1,\ldots,\infty \\ \dot{x} &= f(x)+A\int_0^{r_{max}}\sum_{k=1}^{\infty}c_k(x,a_k,t)\phi_k(r)dr \end{aligned} \tag{9.A.2}$$

while the truncated system of Eq.2.20 takes the form:

$$\begin{aligned} \int_0^{r_{max}} \psi_\nu(r)\sum_{k=1}^{N}\phi_k(r)\frac{\partial a_{kN}(t)}{\partial t}dr &= -G(x_N)\sum_{k=1}^{N}a_{kN}(t)\int_0^{r_{max}}\psi_\nu(r)\frac{\partial \phi_k(r)}{\partial r}dr \\ &\quad +\int_0^{r_{max}}\psi_\nu(r)\sum_{k=1}^{N}b_k(x_N,a_{kN},t)\phi_k(r)dr, \\ &\quad \nu=1,\ldots,N \\ \dot{x}_N &= f(x_N)+A\int_0^{r_{max}}\sum_{k=1}^{N}c_k(x_N,a_{kN},t)\phi_k(r)dr \end{aligned} \tag{9.A.3}$$

Defining the error variables $e_n = n(r,t) - n_N(r,t) = \sum_{k=1}^{\infty}a_k(t)\phi_k(r) - \sum_{k=1}^{N}a_{kN}(t)\phi_k(r)$ and $e_x = x(t) - x_N(t)$ and computing the difference between Eq.9.A.2 and Eq.9.A.3, we get:

$$\begin{aligned}\int_0^{r_{max}} \psi_\nu(r)\frac{\partial e_n}{\partial t}dr &= -G(x)\int_0^{r_{max}} \psi_\nu(r)\frac{\partial n}{\partial r}dr + G(x_N)\int_0^{r_{max}} \psi_\nu(r)\frac{\partial n_N}{\partial r}dr \\ &+ \int_0^{r_{max}} \psi_\nu(r)\sum_{k=1}^{\infty} b_k(x,a_k,t)\phi_k(r)dr \\ &- \int_0^{r_{max}} \psi_\nu(r)\sum_{k=1}^{N} b_k(x_N,a_{kN},t)\phi_k(r)dr \\ \dot{e}_x &= f(x) - f(x_N) + A\int_0^{r_{max}} \sum_{k=1}^{\infty} c_k(x,a_k,t)\phi_k(r)dr \\ &- A\int_0^{r_{max}} \sum_{k=1}^{N} c_k(x_N,a_{kN},t)\phi_k(r)dr\end{aligned} \tag{9.A.4}$$

The above error system is locally exponentially stable (this follows from the assumption that the system of Eq.2.20 is locally exponentially stable for any N), and thus, applying standard results from perturbation theory for finite-dimensional systems (Khalil, 1996), we obtain the following bound for $e_n(r,t), e_x(t)$ for all $t \geq 0$:

$$\begin{aligned} e_n(r,t) &= O(\epsilon(N)) \\ e_x(t) &= O(\epsilon(N)) \end{aligned} \tag{9.A.5}$$

From the above result, the estimates of Eq.2.21 follow directly. △

Proof of Proposition 2.2: To prove that local exponential stability of the system of Eq.2.32 ensures local exponential stability of the system of Eq.2.20, we need to show that the off manifold transients decay exponentially. To this end, we define $e_m = a_{fN} - \tilde{\Sigma}_{\bar{l}+1}(\tilde{x}) = a_{fN} - \tilde{a}_{fN}$ and compute $\dot{a}_{fN}$:

$$\dot{e}_m = \tilde{A}a_{fN} + \tilde{f}_q(a_{fN},\tilde{x}) - \dot{\tilde{a}}_{fN} \tag{9.A.6}$$

From the definition of the approximate inertial manifold $\dot{\tilde{a}}_{fN} \equiv 0 = \tilde{A}\tilde{a}_{fN} + \tilde{f}_q(\tilde{a}_{fN},\tilde{x})$, and thus, the system of Eq.9.A.6 can be written as:

$$\begin{aligned} \dot{e}_m &= \tilde{A}a_{fN} + \tilde{f}_q(a_{fN},\tilde{x}) - \tilde{A}\tilde{a}_{fN} - \tilde{f}_q(\tilde{a}_{fN},\tilde{x}) \\ &= \tilde{A}e_m + \tilde{f}_q(a_{fN},\tilde{x}) - \tilde{f}_q(\tilde{a}_{fN},\tilde{x}) \end{aligned} \tag{9.A.7}$$

Since $\tilde{A}$ is Hurwitz, the terms $\tilde{f}_q(a_{fN},\tilde{x}), \tilde{f}_q(\tilde{a}_{fN},\tilde{x})$ do not include linear terms and the system of Eq.2.32 is locally exponentially stable, one can show that the linearization of the system of Eq.2.26 is exponentially stable, and thus, the nonlinear system is locally exponentially stable.

To establish the closeness of solutions result, we exploit the fact that the local exponential stability implies (Khalil, 1996) $a_{kN}(t) = \tilde{a}_{kN}(t) + \bar{\epsilon}(\bar{l})$, and also that $\Sigma(\tilde{x}) = \tilde{\Sigma}_{\bar{l}+1}(\tilde{x}) + \bar{\epsilon}(\bar{l})$, as $t \to \infty$. Computing the discrepancy $\lim_{t\to\infty} ||n_N - \bar{n}_N||_2$, using the above estimates, it can be shown that $\lim_{t\to\infty} ||n_N - \bar{n}_N||_2 = O(\bar{\epsilon}(\bar{l}))$. △

Proof of Theorem 2.1: Under the controller of Eq.2.39, the closed-loop system takes the form:

$$\begin{aligned}
\frac{d\omega}{dt} &= \tilde{f}(\omega) + \tilde{g}(\omega)\left\{[\beta_{1r_1} \cdots \beta_{mr_m}]C(\omega)\right\}^{-1}\left\{v - \sum_{i=1}^{m}\sum_{k=0}^{r_i}\beta_{ik}L_{\tilde{f}}^{k}\tilde{h}_i(\omega)\right\} \\
&\quad + L(y - \tilde{h}(\omega)) \\
\frac{\partial n}{\partial t} &= -\frac{\partial(G(x,r)n)}{\partial r} + w(n,x,r) \\
\dot{x} &= f(x) + A\int_0^{r_{max}} a(n,r,x)dr \\
&\quad + g(x)\left\{[\beta_{1r_1} \cdots \beta_{mr_m}]C(\omega)\right\}^{-1}\left\{v - \sum_{i=1}^{m}\sum_{k=0}^{r_i}\beta_{ik}L_{\tilde{f}}^{k}\tilde{h}_i(\omega)\right\}
\end{aligned} \tag{9.A.8}$$

Applying the method of weighted residuals to the above system, we obtain the following approximate ODE system:

$$\begin{aligned}
&\frac{d\omega}{dt} = \tilde{f}(\omega) + L(y - \tilde{h}(\omega)) + \tilde{g}(\omega)\left\{[\beta_{1r_1} \cdots \beta_{mr_m}]C(\omega)\right\}^{-1}\left\{v - \sum_{i=1}^{m}\sum_{k=0}^{r_i}\beta_{ik}L_{\tilde{f}}^{k}\tilde{h}_i(\omega)\right\} \\
&\int_0^{r_{max}} \psi_\nu(r)\sum_{k=1}^{N}\phi_k(r)\frac{\partial a_{kN}(t)}{\partial t}dr = -\sum_{k=1}^{N}a_{kN}(t)\int_0^{r_{max}}\psi_\nu(r)\frac{\partial(G(x_N,r)\phi_k(r))}{\partial r}dr \\
&+\int_0^{r_{max}}\psi_\nu(r)w(\sum_{k=1}^{N}a_{kN}(t)\phi_k(r),x_N,r)dr, \quad \nu = 1,\ldots,N \\
&\dot{x}_N = f(x_N) + A\int_0^{r_{max}} a(\sum_{k=1}^{N}a_{kN}(t)\phi_k(r),r,x_N)dr \\
&+g(x_N)\left\{[\beta_{1r_1} \cdots \beta_{mr_m}]C(\omega)\right\}^{-1}\left\{v - \sum_{i=1}^{m}\sum_{k=0}^{r_i}\beta_{ik}L_{\tilde{f}}^{k}\tilde{h}_i(\omega)\right\}
\end{aligned} \tag{9.A.9}$$

Using the vector notation $a_N = [a_{1N} \;\cdots\; a_{NN}]^T$, the above system can be written as:

$$\begin{aligned}
\frac{d\omega}{dt} &= \tilde{f}(\omega) + L(y - \tilde{h}(\omega)) \\
&\quad + \tilde{g}(\omega)\left\{[\beta_{1r_1} \cdots \beta_{mr_m}]C(\omega)\right\}^{-1}\left\{v - \sum_{i=1}^{m}\sum_{k=0}^{r_i}\beta_{ik}L_{\tilde{f}}^{k}\tilde{h}_i(\omega)\right\} \\
\dot{a}_N &= \bar{f}(a_N, x_N) \\
\dot{x}_N &= f(x_N) + A\int_0^{r_{max}} a(\sum_{k=1}^{N}a_{kN}(t)\phi_k(r),x_N,r)dr \\
&\quad + g(x_N)\left\{[\beta_{1r_1} \cdots \beta_{mr_m}]C(\omega)\right\}^{-1}\left\{v - \sum_{i=1}^{m}\sum_{k=0}^{r_i}\beta_{ik}L_{\tilde{f}}^{k}\tilde{h}_i(\omega)\right\}
\end{aligned} \tag{9.A.10}$$

where the explicit form of the nonlinear function $\bar{f}(a_N, x_N)$ is omitted for brevity. Exploiting the orthogonality of the basis functions $\phi_k(r)$ and using the expression for n_N of Eq.2.24, we can write the system of Eq.9.A.10 as:

$$\begin{aligned}
\frac{d\omega}{dt} &= \tilde{f}(\omega) + \tilde{g}(\omega)\left\{[\beta_{1r_1} \cdots \beta_{mr_m}]C(\omega)\right\}^{-1}\left\{v - \sum_{i=1}^{m}\sum_{k=0}^{r_i}\beta_{ik}L_{\tilde{f}}^{k}\tilde{h}_i(\omega)\right\} \\
&\quad + L(y - \tilde{h}(\omega)) \\
\dot{a}_{sN} &= \bar{f}_p(a_{sN}, a_{fN}, x_N) \\
\dot{a}_{fN} &= \bar{f}_q(a_{sN}, a_{fN}, x_N) \\
\dot{x}_N &= f(x_N) + A\int_0^{r_{max}} a(n_p + n_q, x_N, r)dr \\
&\quad + g(x_N)\left\{[\beta_{1r_1} \cdots \beta_{mr_m}]C(\omega)\right\}^{-1}\left\{v - \sum_{i=1}^{m}\sum_{k=0}^{r_i}\beta_{ik}L_{\tilde{f}}^{k}\tilde{h}_i(\omega)\right\}
\end{aligned} \tag{9.A.11}$$

where $\bar{f} = [\bar{f}_p^T \ \bar{f}_q^T]^T$, $a_{sN} = [a_{1N} \cdots a_{pN}]^T$, and $a_{fN} = [a_{(p+1)N} \cdots a_{NN}]^T$. Defining the observer error vector $e_o = \omega - \tilde{x}$, using the assumption $C_L = \frac{1}{\mu}\bar{A}$ where $\bar{A}$ is a Hurwitz matrix, and multiplying the $\dot{e}_o$-subsystem by μ, the system of Eq.9.A.11 can be written as:

$$\begin{aligned}
\mu\frac{de_o}{dt} &= \bar{A}e_o + \mu\hat{f}(\omega, \tilde{x}, v) \\
\dot{a}_{sN} &= \bar{f}_p(a_{sN}, a_{fN}, x_N) \\
\dot{a}_{fN} &= \bar{f}_q(a_{sN}, a_{fN}, x_N) \\
\dot{x}_N &= f(x_N) + A\int_0^{r_{max}} a(n_p + n_q, x_N, r)dr \\
&\quad + g(x_N)\left\{[\beta_{1r_1} \cdots \beta_{mr_m}]C(\omega)\right\}^{-1}\left\{v - \sum_{i=1}^{m}\sum_{k=0}^{r_i}\beta_{ik}L_{\tilde{f}}^{k}\tilde{h}_i(\omega)\right\}
\end{aligned} \tag{9.A.12}$$

where $\hat{f}(\omega, \tilde{x}, v)$ is a nonlinear vector function. The above system is in singularly perturbed form (see Khalil, 1996 for details) and possesses an exponentially stable fast subsystem: $\frac{de_o}{d\tau} = \bar{A}e_o$, where $\tau = t/\mu$, and a slow subsystem that has the form:

$$\begin{aligned}
\dot{a}_{sN} &= \bar{f}_p(a_{sN}, a_{fN}, x_N) \\
\dot{a}_{fN} &= \bar{f}_q(a_{sN}, a_{fN}, x_N) \\
\dot{x}_N &= f(x_N) + A\int_0^{r_{max}} a(n_p + n_q, x_N, r)dr \\
&\quad + g(x_N)\left\{[\beta_{1r_1} \cdots \beta_{mr_m}]C(\tilde{x})\right\}^{-1}\left\{v - \sum_{i=1}^{m}\sum_{k=0}^{r_i}\beta_{ik}L_{\tilde{f}}^{k}\tilde{h}_i(\tilde{x})\right\}
\end{aligned} \tag{9.A.13}$$

Setting $a_{fN} = \tilde{\Sigma}_{\bar{l}+1}(a_{sN}, x_N)$ (model reduction based on the concept of approximate inertial manifold) and using the notation of subsection 3.2.2, we finally obtain:

$$\begin{aligned} \dot{\tilde{x}} &= \tilde{f}(\tilde{x}) + \tilde{g}(\tilde{x}) \{[\beta_{1r_1} \cdots \beta_{mr_m}]C(\tilde{x})\}^{-1} \left\{ v - \sum_{i=1}^{m}\sum_{k=0}^{r_i} \beta_{ik} L_{\tilde{f}}^k \tilde{h}_i(\tilde{x}) \right\} \\ y_{s_i} &= \tilde{h}_i(\tilde{x}), \;\; i = 1, \ldots, m \end{aligned} \tag{9.A.14}$$

Using assumptions 2 and 3 of the theorem, one can show that the above system is locally exponentially stable and $\lim_{t\to\infty} |y_{s_i} - v_i| = 0$. From the assumption that $\bar{l}$ is sufficiently large and Proposition 2.2, we have that the systems of Eq.9.A.13 and Eq.9.A.14 possess the same stability properties, and thus, the slow subsystem of Eq.9.A.13 is also locally exponentially stable. The local exponential stability of the fast and slow subsystems implies (Khalil, 1996) that there exists a μ^* such that if $\mu \in (0, \mu^*]$, then the closed-loop system of Eq.9.A.11 is locally exponentially stable. From the closeness of solutions results of Propositions 2.1 and 2.2, the asymptotic output tracking result $\lim_{t\to\infty} |y_i - v_i| = O(\hat{\epsilon}(N+\bar{l}))$ then follows. △

APPENDIX 9.B: Proof of Chapter 4

Proof of Theorem 4.1 : Substituting the controller of Eq.4.17 into the particulate process model of Eq.4.1, we obtain:

$$\begin{aligned} \frac{\partial n}{\partial t} &= -\frac{\partial (G(x,r)n)}{\partial r} + \bar{w}(n,x,r,z,\theta(t)) + \bar{g}_1(n,x,r)a(\tilde{x},\bar{v},t), \;\; n(0,t) = b(x(t)) \\ \dot{x} &= \bar{f}(x) + Q_1(x)z + \bar{g}_2(x)a(\tilde{x},\bar{v},t) + \bar{g}_3(x,\theta(t), \int_0^{r_{max}} a_2(n,r,x)dr) \\ \epsilon\dot{z} &= \tilde{f}(x) + Q_2(x)z + \tilde{g}_2(x)a(\tilde{x},\bar{v},t) + \tilde{g}_3(x,\theta(t), \int_0^{r_{max}} a_2(n,r,x)dr) \end{aligned} \tag{9.B.1}$$

Performing a decomposition of the above system into the fast and slow time-scales, we obtain the following system that describes the fast dynamics of the closed-loop system:

$$\frac{dz}{d\tau} = \tilde{f}(x) + Q_2(x)z + \tilde{g}_2(x)a(\tilde{x},\bar{v},t) + \tilde{g}_3(x,\theta(t), \int_0^{r_{max}} a_2(n,r,x)dr) \tag{9.B.2}$$

Since the feedback law $a(\tilde{x},\bar{v},t)$ does not use feedback of the fast state z, the above system is exponentially stable. Setting $\epsilon = 0$, the system that describes the slow dynamics of the closed-loop system is obtained:

$$\begin{aligned} \frac{\partial n}{\partial t} &= -\frac{\partial (G(x,r)n)}{\partial r} + w(n,x,r,\theta(t)) + g_1(n,x,r)a(\tilde{x},\bar{v},t), \;\; n(0,t) = b(x(t)) \\ \dot{x} &= f(x) + g_2(x)a(\tilde{x},\bar{v},t) + g_3(x,\theta(t), \int_0^{r_{max}} a_2(n,r,x)dr) \end{aligned} \tag{9.B.3}$$

Applying the method of weighted residuals to the above system, we obtain:

$$\int_0^{r_{max}} \psi_\nu(r)\sum_{k=1}^{N}\phi_k(r)\frac{\partial a_{kN}(t)}{\partial t}dr = -\sum_{k=1}^{N}a_{kN}(t)\int_0^{r_{max}} \psi_\nu(r)\frac{\partial(G(x_N,r)\phi_k(r))}{\partial r}dr$$

$$+\int_0^{r_{max}} \psi_\nu(r)w(\sum_{k=1}^{N}a_{kN}(t)\phi_k(r),x_N,r,\theta(t))dr$$

$$+\int_0^{r_{max}} \psi_\nu(r)g_1(\sum_{k=1}^{N}a_{kN}(t)\phi_k(r),x_N,r)a(\tilde{x},\bar{v},t)dr, \quad \nu=1,\dots,N$$

$$\dot{x}_N = f(x_N)+g_2(x_N)a(\tilde{x},\bar{v},t)+g_3(x_N,\theta(t),\int_0^{r_{max}} a_2(\sum_{k=1}^{N}a_{kN}(t)\phi_k(r),r,x_N)dr)$$

$$y_{s_i}(t) = h_i(\int_0^{r_{max}} c_\kappa\sum_{k=1}^{N}a_{kN}(t)\phi_k(r)dr,x_N), \quad i=1,\dots,m, \quad \kappa=1,\dots,l \tag{9.B.4}$$

or using the vector notation:

$$\begin{aligned}\dot{\tilde{x}} &= \tilde{f}(\tilde{x})+\sum_{i=1}^{m}\tilde{g}_i(\tilde{x})a_i(\tilde{x},\bar{v},t)+\tilde{w}(\tilde{x},\theta)\\ y_{s_i} &= \tilde{h}_i(\tilde{x}), \quad i=1,\dots,m\end{aligned} \tag{9.B.5}$$

For the above system we have shown in Christofides, 1998 that there exists a $\phi\in(0,\phi^*]$ such that if $\max\{|\tilde{x}(0)|,||\theta||,||\dot{\theta}||,||\bar{v}||\}\leq\delta$, then its state is bounded and its outputs satisfy $\limsup_{t\to\infty}|y_{s_i}(t)-v_i|\leq O(\phi)$, $i=1,\dots,l$. Finally, since the fast subsystem of Eq.9.B.2 is exponentially stable, we can use standard stability results for singularly perturbed systems to obtain that if $\max\{|x(0)|,||\eta(0)||_2,||\theta||,||\dot{\theta}||,||\bar{v}||\}\leq\delta$, then the state of the closed-loop system is bounded and that its outputs satisfy the relation of Eq.4.18. △

APPENDIX 9.C: Proofs of Chapter 5

Proof of Theorem 5.1 : Consider the representation of the system of Eq.5.12 in the form of Eqs.5.20-5.21. Under the controller of Eq.5.28, the closed-loop system takes the following form:

$$\begin{aligned}\frac{d\omega}{dt} &= \tilde{f}(\omega)-\frac{1}{2}\tilde{g}(\omega)R^{-1}(\omega)I\left(L_{\tilde{g}}V(\omega)\right)^T+L(y-\tilde{h}(\omega))\\ \dot{e} &= Ae+B\left[l_1(e,\eta,\bar{v})-\frac{1}{2}C(\tilde{x})R^{-1}(\omega)I\left(L_{\tilde{g}}V(\omega)\right)^T\right]\\ \dot{\eta}_1 &= \Psi_1(\zeta,\eta)\\ &\vdots\\ \dot{\eta}_{(n+N)-\sum_i r_i} &= \Psi_{(n+N)-\sum_i r_i}(\zeta,\eta)\\ y_{s_i} &= \zeta_1^{(i)}, \quad i=1,\dots,m\end{aligned} \tag{9.C.1}$$

Defining the observer error vector $e_o = \omega - \tilde{x}$, using the assumption $C_L = \frac{1}{\mu}\bar{A}$ where $\bar{A}$ is a Hurwitz matrix, and multiplying the $\dot{e}_o$-subsystem by μ, the system of Eq.9.C.1 can be written as:

$$
\begin{aligned}
\mu \frac{de_o}{dt} &= \bar{A}e_o + \mu \hat{f}(\omega, \tilde{x}, v) \\
\dot{e} &= Ae + B\left[l_1(e,\eta,\bar{v}) - \frac{1}{2}C(\tilde{x})R^{-1}(\omega)I\left(L_{\bar{g}}V(\omega)\right)^T\right] \\
\dot{\eta}_1 &= \Psi_1(\zeta,\eta) \\
&\vdots \\
\dot{\eta}_{(n+N)-\sum_i r_i} &= \Psi_{(n+N)-\sum_i r_i}(\zeta,\eta) \\
y_{s_i} &= \zeta_1^{(i)}, \quad i = 1,\ldots,m
\end{aligned}
\tag{9.C.2}
$$

where $\hat{f}(\omega, \tilde{x}, v)$ is a nonlinear vector function. The above system is in singularly perturbed form (see Khalil, 1996 for details) and possesses an exponentially stable fast subsystem: $\frac{de_o}{d\tau} = \bar{A}e_o$, where $\tau = t/\mu$, and a slow subsystem that has the form:

$$
\begin{aligned}
\dot{e} &= Ae + B\left[l_1(e,\eta,\bar{v}) - \frac{1}{2}C(\tilde{x})R^{-1}(\tilde{x})I\left(L_{\bar{g}}V(\tilde{x})\right)^T\right] \\
\dot{\eta}_1 &= \Psi_1(\zeta,\eta) \\
&\vdots \\
\dot{\eta}_{(n+N)-\sum_i r_i} &= \Psi_{(n+N)-\sum_i r_i}(\zeta,\eta) \\
y_{s_i} &= \zeta_1^{(i)}, \quad i = 1,\ldots,m
\end{aligned}
\tag{9.C.3}
$$

Consider now the Lyapunov function candidate $V = e^T P e$ introduced in theorem 5.1. Evaluating the time-derivative of this function along the trajectories of the closed-loop slow e-subsystem in Eq.9.C.3, we obtain:

$$
\begin{aligned}
\dot{V} &= L_{\bar{f}}V + L_{\bar{g}}Vu \\
&= L_{\bar{f}}V - L_{\bar{g}}V \frac{L_{\bar{f}}^*V + \sqrt{(L_{\bar{f}}^*V)^2 + (u_{max}^2(L_{\bar{g}}V)(L_{\bar{g}}V)^T)^2}}{(L_{\bar{g}}V)(L_{\bar{g}}V)^T\left[1 + \sqrt{1 + u_{max}^2(L_{\bar{g}}V)(L_{\bar{g}}V)^T}\right]}(L_{\bar{g}}V)^T \\
&= \frac{-\rho|e|^2 + L_{\bar{f}}V\sqrt{1 + u_{max}^2(L_{\bar{g}}V)(L_{\bar{g}}V)^T} - \sqrt{(L_{\bar{f}}^*V)^2 + (u_{max}^2(L_{\bar{g}}V)(L_{\bar{g}}V)^T)^2}}{\left[1 + \sqrt{1 + u_{max}^2(L_{\bar{g}}V)(L_{\bar{g}}V)^T}\right]}
\end{aligned}
\tag{9.C.4}
$$

It is clear from the above equation and the fact that $\rho > 0$ that whenever $L_{\bar{f}}^*V \leq 0$, we have $L_{\bar{f}}V \leq 0$ and the time-derivative of V satisfies:

$$
\dot{V} \leq \frac{-\rho|e|^2}{\left[1 + \sqrt{1 + u_{max}^2(L_{\bar{g}}V)(L_{\bar{g}}V)^T}\right]} < 0 \quad \forall \ e \neq 0 \tag{9.C.5}
$$

Furthermore, whenever $0 < L_{\bar{f}}^*V \leq u_{max}|L_{\bar{g}}V|$, we have $\sqrt{(L_{\bar{f}}^*V)^2 + (u_{max}^2(L_{\bar{g}}V)(L_{\bar{g}}V)^T)^2} \geq L_{\bar{f}}V\sqrt{1 + u_{max}^2(L_{\bar{g}}V)(L_{\bar{g}}V)^T}$ and $\dot{V}$ satisfies

Eq.9.C.5. Summarizing, we have that whenever $L_{\bar{f}}^{*}V \leq u_{max}|L_{\bar{g}}V|$, the closed-loop slow e-subsystem is asymptotically stable. Therefore, given any positive real number δ_x such that $|\tilde{x}(0)| \leq \delta_x$ (and Eq.5.27 holds for all $|\tilde{x}| \leq \delta_x$), there exists a positive real number M such that $|\tilde{x}(t)| \leq M \ \forall \ t \geq 0$ (i.e., the closed-loop trajectories are bounded) and, consequently, the denominator expression in Eq.9.C.5 is bounded. It follows then that there exists a positive real number k_1 such that $\dot{V} \leq -k_1|e|^2$ and the origin of the closed-loop slow e-subsystem is exponentially stable. Therefore, the following inequality holds for the e-states of the closed-loop slow system:

$$|e(t)| \quad \leq \quad K_e|e(0)|e^{-a_1 t} \quad \forall \ t \geq 0 \tag{9.C.6}$$

for some $K_e \geq 1, a_1 > 0$. From assumption 2, we have that the η states of the closed-loop slow system satisfy the following inequality:

$$|\eta(t)| \quad \leq \quad K_\eta|\eta(0)|e^{-at} + \gamma_\eta(||e||) \quad \forall \ t \geq 0 \tag{9.C.7}$$

for some $K_\eta \geq 1, a > 0$, where γ_η is a class K function . At this point, one can show, using the results in Christofides and Teel, 1996, that the inequalities of Eqs.9.C.6-9.C.7 continue to hold, for the states of the closed-loop singularly perturbed system of Eq.9.C.2, up to an arbitrarily small offset, d, for initial conditions in arbitrarily large compact subsets of the region described by Eq.5.27, provided that the singular perturbation parameter μ is sufficiently small. This offset can be selected to be sufficiently small such that, after a sufficiently large time (when the exponential terms in Eqs.9.C.6-9.C.7 die out), the states of the closed-loop system are confined within a small ball centered around the origin. Direct application then of the result of theorem 9.3 in Khalil, 1996 yields that there exists a μ^* such that if $\mu \leq \mu^*$, the closed-loop system of Eq.9.C.2 is exponentially stable (A somewhat similar argument was used in the proof of theorem 1 in El-Farra and Christofides, 2001b, see this reference for mathematical details.) It follows then that:

$$|e(t)| \quad \leq \quad \alpha|e(0)|e^{-\beta t} \quad \forall \ t \geq 0 \tag{9.C.8}$$

for some $\alpha \geq 1$ and $\beta > 0$. The asymptotic output tracking result can be easily obtained by taking the limit of both sides of Eq.9.C.8 as $t \to \infty$

$$\lim_{t\to\infty} |e_i(t)| = \lim_{t\to\infty} |y_{si}(t) - v_i(t)| = 0, \quad i = 1, \cdots, m \tag{9.C.9}$$

△

Proof of Theorem 5.2 : Under the controller of Eq.5.28, the infinite-dimensional closed-loop system takes the form:

$$\begin{aligned}
\frac{d\omega}{dt} &= \tilde{f}(\omega) - \frac{1}{2}\tilde{g}(\omega)R^{-1}(\omega)(L_{\bar{g}}V(\omega))^T + L(y - \tilde{h}(\omega)) \\
\frac{\partial n}{\partial t} &= -\frac{\partial(G(x,r)n)}{\partial r} + w(n,x,r) \\
\dot{x} &= f(x) - \frac{1}{2}g(x)R^{-1}(\omega)(L_{\bar{g}}V(\omega))^T + A\int_0^{r_{max}} a(n,r,x)dr
\end{aligned} \tag{9.C.10}$$

Applying the method of weighted residuals to the above system, we obtain the following approximate ODE system:

$$\frac{d\omega}{dt} = \tilde{f}(\omega) - \frac{1}{2}\tilde{g}(\omega)R^{-1}(\omega)(L_{\tilde{g}}V(\omega))^T + L(y - \tilde{h}(\omega))$$

$$\int_0^{r_{max}} \psi_\nu(r)\sum_{k=1}^{N}\phi_k(r)\frac{\partial a_{kN}(t)}{\partial t}dr = -\sum_{k=1}^{N}a_{kN}(t)\int_0^{r_{max}}\psi_\nu(r)\frac{\partial(G(x_N,r)\phi_k(r))}{\partial r}dr$$

$$+\int_0^{r_{max}}\psi_\nu(r)w(\sum_{k=1}^{N}a_{kN}(t)\phi_k(r),x_N,r)dr,\;\; \nu = 1,\dots,N$$

$$\dot{x}_N = f(x_N) - \frac{1}{2}g(x_N)R^{-1}(\omega)(L_{\tilde{g}}V(\omega))^T + A\int_0^{r_{max}} a(\sum_{k=1}^{N}a_{kN}(t)\phi_k(r),r,x_N)dr \tag{9.C.11}$$

Using the vector notation $a_N = [a_{1N}\ \cdots\ a_{NN}]$, and after some rearrangements, the above system can be written as:

$$\begin{aligned}
\frac{d\omega}{dt} &= \tilde{f}(\omega) - \frac{1}{2}\tilde{g}(\omega)R^{-1}(\omega)(L_{\tilde{g}}V(\omega))^T + L(y - \tilde{h}(\omega)) \\
\dot{a}_N &= f^*(a_N, x_N) \\
\dot{x}_N &= f(x_N) - \frac{1}{2}g(x_N)R^{-1}(\omega)I(L_{\tilde{g}}V(\omega))^T \\
&\quad +A\int_0^{r_{max}} a_2(\sum_{k=1}^{N}a_{kN}(t)\phi_k(r),r,x_N)dr)
\end{aligned} \tag{9.C.12}$$

where the explicit expression of the nonlinear function $f^*(a_N, x_N)$ is omitted for brevity. Setting $\tilde{x} = [a_N^T\ x_N^T]^T$, we finally obtain the following closed-loop system:

$$\begin{aligned}
\frac{d\omega}{dt} &= \tilde{f}(\omega) - \frac{1}{2}\tilde{g}(\omega)R^{-1}(\omega)(L_{\tilde{g}}V(\omega))^T + L(y - \tilde{h}(\omega)) \\
\dot{\tilde{x}} &= \tilde{f}(\tilde{x}) - \frac{1}{2}\sum_{i=1}^{m}\tilde{g}_i(\tilde{x})R^{-1}(\omega)L_{\tilde{g}_i}V(\omega),\;\; i = 1,\dots,m
\end{aligned} \tag{9.C.13}$$

Under assumptions 1, 2, and 3 stated in theorem 5.1, we have already shown in the proof of theorem 5.1 above that there exists $\mu^* > 0$ such that if $\mu \in (0, \mu^*]$, the above system is locally exponentially stable and $\lim_{t\to\infty}|y_{si} - v_i| = 0$. Using the results of proposition 2.1, we have that, for sufficiently large N, the following estimates hold

$$\begin{aligned}
n(r,t) &= n_N(r,t) + O(\epsilon(N)) \\
x(t) &= x_N(t) + O(\epsilon(N))
\end{aligned} \tag{9.C.14}$$

where $\epsilon(N)$ is a small positive real number that depends on N and satisfies $\lim_{N\to\infty}\epsilon(N) = 0$, and $n_N(r,t) = \sum_{k=1}^{N} a_{kN}(t)\phi_k(r)$ is the approximation of $n(r,t)$. It follows then that, for sufficiently large N, there exist positive real numbers $\delta_n, \bar{\delta}_x, \delta_\omega, \bar{\mu}^*$ such that if $\mu \in (0, \bar{\mu}^*]$, $||n(r,0)||_2 < \delta_n$, $|x(0)| < \bar{\delta}_x$, $|\omega(0)| < \delta_\omega$, the closed-loop system of Eq.9.C.10 is exponentially stable and $\lim_{t\to\infty}|y_i - v_i| = O(\epsilon(N))$.

APPENDIX 9.D: INDEX

PARTICLE TECHNOLOGY SERIES

FORMERLY POWDER TECHNOLOGY SERIES

1. H. Rumpf (Translated by F.A. Bull): *Particle Technology*. 1990 ISBN 0-412-35230-3
2. S. Lowell and Joan E. Shields: *Powder Surface Area and Porosity* (Third Edition). 1991 ISBN 0-412-39690-4
3. K. Heiskanen: *Particle Classification*. 1993 ISBN 0-412-49300-4
4. O. Molerus: *Principles of Flow in Disperse Systems*. 1993 ISBN 0-412-40630-6
5. I. Claus Bernhardt: *Particle Size Analysis*. Classification and Sedimentation Methods. 1994 ISBN 0-412-55880-7
6. Yasuo Arai: *Chemistry of Powder Production*. 1996 ISBN 0-412-39540-1
7. Terence Allen: *Particle Size Measurement* (Fifth Edition). 1996
   Volume 1: Surface area and pore size determination. ISBN 0-412-72950-4
   Volume 2: Powder sampling and particle size measurement. ISBN 0-412-75330-8
   Set of 2 vols., ISBN 0-412-75350-2
8. George E. Klinzing, Roy D. Marcus, Farid Rizk and L.S. Leung†: *Pneumatic Conveying of Solids*. A Theoretical and Practical Approach (Second Edition). 1997 ISBN 0-412-72440-5
9. Jonathan P.K. Seville, Uğur Tüzün and Roland Clift: *Processing of Particulate Solids*. 1997 ISBN 0-7514-0376-8
10. B. Kaye: *Powder Mixing*. 1997 ISBN 0-412-40340-4
11. O. Molerus and K.-E. Wirth: *Heat Transfer in Fluidized Beds*. 1997 ISBN 0-412-60800-6
12. R. Weinekötter and H. Gericke: *Mixing of Solids*. 2000 ISBN 0-7923-6229-2
13. R. Xu: *Particle Characterization: Light Scattering Methods*. 2000 ISBN 0-7923-6300-0
14. P.D. Christofides: *Model-Based Control of Particulate Processes*. 2002 ISBN 1-4020-0936-4

KLUWER ACADEMIC PUBLISHERS – DORDRECHT / BOSTON / LONDON